Terahertz, Ultrafast Lasers and Their Medical and Industrial Applications

Edited by Sulaiman Wadi Harun

Published in London, United Kingdom

Terahertz, Ultrafast Lasers and Their Medical and Industrial Applications
http://dx.doi.org/10.5772/intechopen.100712
Edited by Sulaiman Wadi Harun

Contributors
Li Wang, Tatsuhiko Aizawa, Tadahiko Inohara, Yohei Suzuki, Tomomi Shiratori, Mehra S. Sidhu, Nitish Dhingra, Sana Niazi, Farideh Doroodgar, Sumit Kumar

First published in London, United Kingdom, 2022 by IntechOpen
IntechOpen is the global imprint of INTECHOPEN LIMITED, registered in England and Wales, registration number: 11086078, 5 Princes Gate Court, London, SW7 2QJ, United Kingdom

British Library Cataloguing-in-Publication Data
A catalogue record for this book is available from the British Library

Additional hard and PDF copies can be obtained from orders@intechopen.com

Terahertz, Ultrafast Lasers and Their Medical and Industrial Applications
Edited by Sulaiman Wadi Harun
p. cm.
Print ISBN 978-1-80356-263-6
Online ISBN 978-1-80356-264-3
eBook (PDF) ISBN 978-1-80356-265-0

Meet the editor

Sulaiman Wadi Harun is a Professor of Photonics at the Department of Electrical Engineering, University of Malaya, Malaysia. He has contributed significantly to the field of fiber-optic and laser technology. He has nearly 20 years of research experience in the development of optical fiber devices including fiber amplifiers, fiber lasers, and fiber optic sensors. Prof. Harun is a fellow of the Malaysian Academy of Science and the founder and honorary advisor for the Optical Society of Malaysia. He received his Ph.D. in Photonics Technology from the University of Malaya in 2004. Prof. Harun received the Malaysian Rising Award in 2016 from the Malaysian Ministry of Higher Education for his contribution to international collaboration.

Contents

Preface

Lasers enable the modification of living and non-living matter with submicron precision in a contact-free manner. They are wonderful devices with enormous application potential in medicine and industry. This book presents the latest research in laser technology.

Terahertz quantum cascade lasers (THz-QCLs) have attracted tremendous interest in recent years due to their coherent and compact wave features. They can be developed using semiconductor quantum structures and the THz radiation obtained is attributed to inter-sub-band transitions, which rely on quantum transport between discrete sub-bands. Electron injection plays a significant role to achieve population inversion in a QCL. Chapter 1 investigates the effect of an indirect injection scheme in terahertz QCL designs. For the scattering-assisted (SA) injection method-based QCLs with two wells, it is found that the population inversion increases when additional high-lying sub-bands are included, owing to the activation of more depopulation channels. However, an increase in the optical gain does not correspondingly occur if the lasing frequency exceeds 3 THz; instead, the peak gain undergoes a significant decrease. This finding indicates that the high-lying sub-bands play an important role in the development of an efficient SA-QCL design. THz-QCLs can be used for many medical applications including imaging, biochemical label-free sensing, pharmacology, and more.

The continuous wave (CW) laser has been around since 1960 and, as its name suggests, it emits a constant laser beam. The advancement in the laser technology has given birth to pulsed lasers, which emit short 'pulses' of light at regular intervals, rather than continuously. For instance, femtosecond lasers release extremely brief pulses of light at time periods on the order of one quadrillionth of a second. One of the key advantages of femtosecond lasers is that they can generate huge amounts of peak power, while also massively reducing the heat produced during this process. Chapter 2 describes the basic principles of a femtosecond laser and its application in ophthalmology. Since femtosecond lasers produce lower levels of heat compared to CW or other longer pulsed lasers, they are suitable for implementing various ophthalmic procedures, such as corneal surgery, laser eye or 'refractive' surgery, and operations to remove cataracts from the eye. This is because minimal heat is especially important when operating on sensitive areas like the eye to prevent damage to the surrounding tissue. Femtosecond lasers promote safe surgery and fast healing times during ophthalmic procedures because they can process tissue materials within a 3D volume without altering its surface.

Pulsed lasers can also be used in other medical applications such as photo-thermal therapy for cancer treatment. A short-pulsed laser is generally used in laser-based photo-thermal therapy to destroy cancerous cells. The major challenge in this therapy is to destroy the cancerous cells without damaging the surrounding healthy tissue. Thus, it is essential to understand the thermal characteristics of the laser-irradiated biological tissue to improve the efficacy of laser-based photo-thermal therapy. Chapter 3 discusses the modelling of laser-irradiated biological tissue to understand its thermal behavior, which may help to improve the efficacy of laser-based photo-thermal therapy. In this study, the light propagation through the biological tissue is mathematically modelled using the heat transfer equation (RTE). RTE is solved using the discrete ordinate method (DOM) to determine the intensity inside the laser-irradiated biological tissue. Consequently, the absorbed photon energy acts as the source term in the Fourier/non-Fourier model-based bio-heat transfer equation to determine the temperature distribution inside the biological tissue subjected to short-pulse laser irradiation.

Laser surface texturing is a top-down method for generating surface patterns on polymers, metals, ceramics, glasses, and alloys. This technique allows large-scale surface patterning. Chapter 4 describes the use of a femtosecond laser in micro-/nano-texturing for fabricating coated and surface-treated dies with tailored textures. Through the femtosecond laser micro-/nano-texturing and CNC-imprinting, the metal, polymer, and glass product surfaces were optically decorated to have color grading and plasmonic brilliance and functionally controlled to be hydrophobic. The proposed approach can be used for micro-/nano-texturing of various industrial and medical products.

The interactions of concentrated energy fluxes such as femtosecond lasers and high-energy electron beams with absorbing substances have facilitated new discoveries and excitement in various scientific and technological areas. For instance, femtosecond laser ablation is an effective technique to functionalize surfaces. Due to the ultrashort pulse width and high light intensity (1012 W/cm2), it is possible for the laser to ablate or irreversibly modify the materials with negligible damage outside the focal volume, thereby allowing treatment of biological samples like live cells, membranes, and removal of thin films as well as bulk materials for many applications in diverse fields including micro-optics, electronics, and even biology under extremely high precision. Chapter 5 discusses the ablation of materials using femtosecond lasers and electron beams. Both femtosecond laser and e-beam ablations are being investigated for a range of medical applications.

This book is written by experts in the field and is a useful resource for researchers, engineers, and advanced students in the field of photonics, lasers, ultrafast optics, material processing, and medical physics.

Sulaiman Wadi Harun
Department of Electrical Engineering,
University of Malaya,
Kuala Lumpur, Malaysia

Chapter 1

High-Lying Confined Subbands in Terahertz Quantum Cascade Lasers

Li Wang

Abstract

In designing the terahertz quantum cascade lasers, electron injection manner indeed plays a significant role to achieve the population inversion. The resonant tunneling process is commonly employed for this injection process but waste more than 50% fraction of populations out of the active region owing to resonance alignment, and the injection efficiency is obviously degraded due to thermal incoherence. An alternative approach is to consider the phonon-assisted injection process that basically contributes to most of the populations to the upper lasing level. However, this manner is still not realized in experiments if a short-period design only containing two quantum wells is used. In this work, it is found in this design that the population inversion is indeed well improved; however, the optical gain is inherently low even at a low temperature. Those two opposite trends are ascribed to a strong parasitic absorption overlapping the gain. The magnitude of this overlap is closely related to the lasing frequency, where frequencies below 3 THz suffer from fewer effects.

Keywords: intersubband transition, terahertz, quantum cascade lasers, parasitic channels, optical gain

1. Introduction

Thus far, the profusion of terahertz wave applications, including high-speed communications, industrial quality control, non-destructive cross-sectional imaging, gas and pollution sensing, biochemical label-free sensing, pharmacology, and security screening, has been demonstrated [1–3]. Moreover, the development of terahertz quantum cascade lasers (THz-QCLs) based on semiconductor quantum structures affords an attractive THz radiation source with coherent and compact wave features [4]. The basic radiation mechanism in this type of laser is intersubband transitions relying on quantum transport between discrete subbands. This method prevents the semiconductor bandgap limit at significantly low THz photon energies. The subbands can be freely tailored via engineering the thickness of quantum layers; therefore, the THz radiation frequency coverage is broad. However, THz-QCLs always suffer from temperature-triggered lasing quenching; consequently, the maximum operating temperature (T_{max}) is still limited to below room temperature and thus requires additional cooling. Notably, despite stable progress since the first reported THz-QCL operating at 50 K (4.4 THz, pulsed mode) [5], T_{max} has been stalled since 2012 (199.5 K) [6]. A recent breakthrough, achieving 250 K operation [7], indeed soothes the uncertainties regarding whether a 300-K-operation is prevented by any physical limit. The result has since spurred further efforts to achieve room-temperature operation.

With regard to the high-temperature THz-QCLs designs, different theoretical models, including density matrix formalism [8–10], non-equilibrium Green's function (NEGF) [11–13], and Monte Carlo techniques [14, 15], have been proposed to understand the effects of temperature on quantum transport, that is, the loss of coherence, parasitic tunneling channels, and non-radiative processes with an increase in temperature. Numerous designs have been proposed, for example, by using diagonal radiative transitions [3] to suppress the thermally activated non-radiation channels (that triggers longitudinal optical (LO)-phonon emission instead of photon emission) between the upper and lower laser subbands, by using phonon resonance to depopulate the lower laser subband [6] yielding higher population inversion and partially relaxing the thermal backfilling, or by using clean subband systems [7] to avoid perturbation from high-lying subbands. Most of these designs use the resonant tunneling (RT) injection mechanism to populate the upper laser subband. In fact, the core feature of QCL design is electrons cascading across hundreds of stacked radiation periods. Therefore, a critical innovation in QCL design is the development of an injector region to maintain stable electrical bias in operations, enabling the first successfully operation of QCLs in the mid-infrared range [4].

However, RT injection in THz-QCLs has several drawbacks, which are illustrated in **Figure 1a**. 1) *Resonance alignment between the injector and upper laser subbands.* Electrons always wait for resonance before being injected into the upper laser subband ($i \rightarrow u$). Ideally, in the coherent transport regime, the upper laser subband u holds as many carriers as the injector subband i, which is half of the total available electrons (that means a maximum of 50% share in the upper laser subband). In reality, with a thick injector barrier and the presence of multiple scattering channels, the population inversion generally falls below 50% at low temperatures. 2) *Thermal backfilling.* Moreover, as the injector subband i is already mostly populated, thermal backfilling to the lower laser subband l cannot be neglected at high temperatures [16, 17]. This process

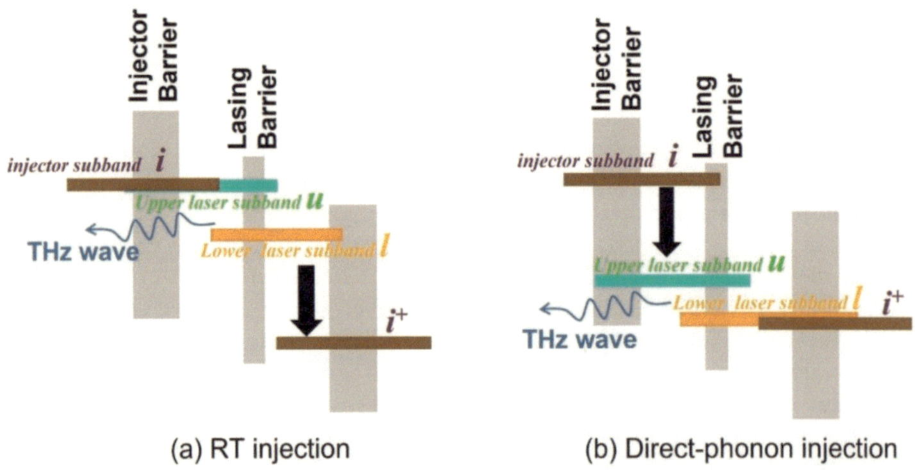

Figure 1.
Illustration of resonant tunneling (RT) injection (a) and phonon assisted injection (direct-phonon injection) (b).

introduces twice the adverse impacts, reducing the population inversion by decreasing the injected population and refilling the lower laser subband l simultaneously. Given that the population inversion is defined as the difference between the upper and lower laser subband populations, the inversion will undergo significant degradation. 3) *Selective injection issue*. Owing to the small subband energy separation, THz-QCLs face difficulty in selectively injecting electrons into the upper laser subband, while avoiding incorrect injection into the lower laser subband. To enhance this selectivity, the injector barrier (**Figure 1a**) should be chosen meticulously, on one hand, the barrier needs to be sufficiently thick to suppress incorrect injection and reduce the negative differential resistance (NDR) during parasitic alignment bias (i.e. the injector subband i and lower laser subband l will first align ($i \leftrightarrow l$) before reaching the operational bias). However, on the other hand, this barrier must also be sufficiently thin to increase the dynamic range of the laser current density. It is important to note that, owing to the close energy spacing between subbands u and l, the subband alignments $i \leftrightarrow u$ and $i \leftrightarrow l$ occur with similar electric fields. The constraint on the injector barrier worsens when the device lasing frequency approaches small. All the aforementioned RT-QCL issues result in significant challenges in terms of maintaining the population inversion when the operating temperature approaches 300 K, thus impelling designers to develop novel approaches to overcome the bottlenecks associated with RT injection.

An indirect injection scheme (scattering-assited (SA) injection), by designing the injector subband i to lie one LO-phonon energy above the upper laser subband u (**Figure 1b**), has also been reported [16, 18–20]. SA injection can be traced back as early as 2001 by Scamarcio *et al.* [21], where the method was successfully employed in mid-infrared QCLs [17] to inject electrons into the upper laser subband by resonantly emitting phonons from the injector subband ($i \rightarrow u$). In this case, by combining a diagonal radiative transition, the parasitic off-resonant tunneling from the injector subband i to the lower laser subband l also can be suppressed. Therefore, this scheme may offer a venue for realizing higher population inversion, even at 300 K. Previous reports [16, 17, 19, 21] provide comprehensive discussions on how SA-QCL designs address the shortcomings of RT-QCLs. However, only three-well [22], four-well [16], and five-well [19] SA-QCL designs have been studied. Moreover, in high-temperature THz-QCLs designs, it is found that a higher $T_{\max}$ also results from a shorter period length by decreasing the number of desired subbands (seven-well [23], four-well [24], three-well [6], and two-well [25]). In this work, we attempt to study the SA injection in THz-QCL designs based on a short period length that contains only two wells. The result demonstrates a strong deviation between the optical gain and population inversion. In other words, although high population inversion can remain in such designs, when the lasing frequency exceeds 3 THz, the peak gain is considerably small even at low temperatures. This phenomenon is ascribed to the emergence of specific parasitic absorption, which can closely overlap with the optical gain. However, this limitation can be avoided in low-frequency lasing. This implies that specific strategies are required to alleviate the shortcomings of different lasing frequencies when introducing short-period SA injection designs; here in the final part of this work, we also study the feasibility in using step well quantum structures to suppress the absorption effects on the gain.

2. Non-equilibrium Green's function method

The most fundamental tool in the design of THz-QCLs structures and analysis is a numerical package to calculate subband wavefunctions and energies. Because the subband energy position is critical for the discussion of the parasitic absorption in this work, it needs to estimate the high-lying energy position more precisely. Here, two factors effecting the energy separation between subbands are considered, **a.** In THz-QCLs, the quantum structure contains the layers with thickness of only several nanometers; the non-parabolicity can largely effect on the energy of confined subbands, especially on HCS as it is lifted further away from the bottom of the conduction band; and the high-electric filed operation of THz-QCLs will worse this issue. Here, the band structures are based on three-band Hamiltonian that accounts for the conduction (*c*), light-hole (*lh*), and split-off (*so*) bands, as follows:

$$H = \begin{pmatrix} E_c(z) + S(z)\dfrac{\hbar^2 k_z^2}{2m_0} & i\sqrt{\dfrac{2}{3}}P(z)k_z & -i\sqrt{\dfrac{1}{3}}P(z)k_z \\ -i\sqrt{\dfrac{2}{3}}P(z)k_z & E_{lh}(z) + (1+L(z))\dfrac{\hbar^2 k_z^2}{2m_0} & 0 \\ i\sqrt{\dfrac{1}{3}}P(z)k_z & 0 & E_{so}(z) + (1+L(z))\dfrac{\hbar^2 k_z^2}{2m_0}(z) \end{pmatrix} \quad (1)$$

where P is the interband momentum matrix element related to the Kane energy E_p through: $P(z) = \sqrt{\frac{m_0 E_p(z)}{2}}$. By comparing the three-band and one-band models for calculating the high-lying subband energy, it finds an approximate 2 meV difference. **b.** the alignment of subbands and also the energy position is also very sensitive to the conduction band offset (CBO) values, especially in short period with tall barriers (here, AlAs% of AlGaAs barrier is 0.3). We follow the latest calibration of CBO based on a machine learning method reported in Ref. [7]. The exact numbers of subbands participating in the transports were controlled by the axial cut-off energy. These subbands were transformed into localized basis modes (reduced real space basis) and used in the NEGF algorithm [26, 27]. The subband energy broadening can play significant roles for estimating the tunneling current (increase the dephasing) and the optical gain (widen the radiation linewidth), especially for THz-QCL studied in this work, as this subband energy broadening (~10 meV) is similar as the photon energy (15 meV). In THz-QCLs, this broadening originates from multiple scattering couplings. Here, the self-energy terms are calculated for all scatterings, including the optical phonons, acoustic phonons, charged impurities, interface roughness, alloy disorder, and electron-electron interactions [28–32]. The critical part of the model is a self-consistent NEGF solver that starts from an initial guess of the Green's functions, the self-energies are then presented roughly, and the Green's functions are again calculated iteratively. Simultaneously, the mean field electrostatic potential is calculated self-consistently (Poisson's equation). Such iterations are performed until convergence is reached. The current density as well as the carrier density distribution is finally displayed. The optical gain or absorption in pairs of intersubband transitions follows linear response theory.

3. Results

A direct comparison of the population residual at individual desired subbands under different injection methods (RT and SA injections) is presented in **Figure 2**. For simplicity, both the designs are based on the two-well quantum structure with a lasing

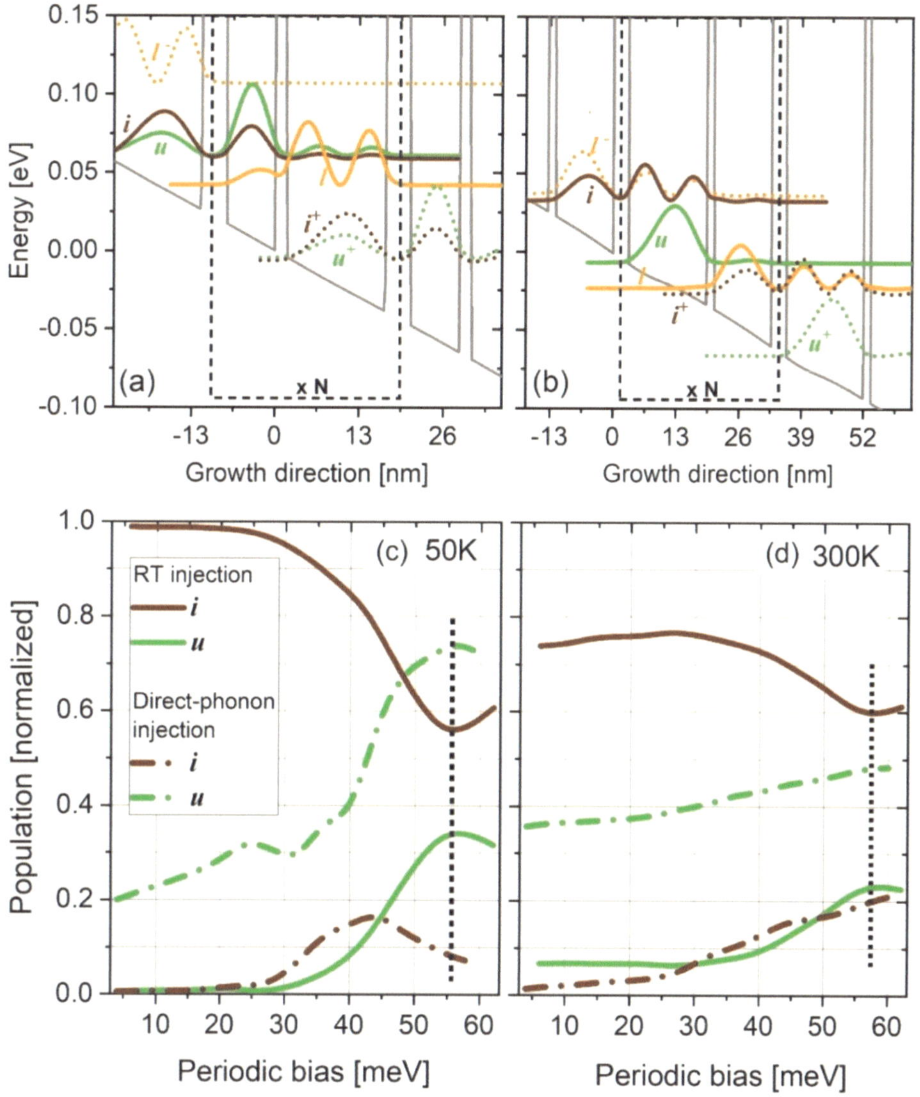

Figure 2.
Design structures with RT injection (a) and SA injection (b). The designs are based on simple quantum structures containing only two wells (three desired subbands). The population shares of the injector subband and the upper laser subband at both low/high temperatures ((c) 50 K and (d) 300 K) are shown to illustrate the advantage of SA injection under the different applied biases (the black dash lines represent the operational bias condition).

frequency of 3.8 THz. The RT-QCL design in **Figure 2a** precisely follows the previously used scheme for $T_{max} > 200$ K [7, 25]. The design in **Figure 2b**, i.e. the scheme designed in this study, employs SA injection. In the former design, electrons are pumped from the injector subband i into the upper laser subband u via RT process. Subsequently, diagonal radiation transition occurs between the laser subbands u and i. Electrons are then depopulated via intrawell LO phonon (direct-phonon) resonance and moved into the next injector subband to repeat the previous steps. In the latter design, electrons are injected from subband i following direct-phonon resonance (phonon emission in a vertical manner) and then perform diagonal transition radiation, after which the depopulation of subband l follows an RT process. It is clear from **Figure 2c** and **d** that, for RT injection, because the bias is applied until it reaches the operational bias (dashed line labeled in **Figure 2c** and **d**), most of the electrons are residual at injector subband i. Under the operational bias, the injector subband i population ratios are 56%/60% at 50 K/300 K. By contrast, the upper laser subband u only maintains 33%/22% at 50 K/300 K. For SA injection, most of the population is injected into the upper laser subband u (72%/58% at 50 K/300 K under the operational bias). Here, the total electrons population in each period is normalized as 100%.

The two-well SA-QCL designs are shown in **Figure 3** with the different lasing frequencies of 2.2 THz (a, b) and 3.8 THz (c, d). To study the effect of high-lying subbands, the number of confined subbands in each period is controlled by tuning the axial cut-off energy range, that is, the narrow range in **Figure 3a** and **c** which only contains three desired subbands (i, u, and l), and the large range in **Figure 3b** and **d** which contains one more high-lying subband together with the desired subbands (i, u, l, and h). To further enhance injection selectivity, the lasing barrier positioned between the upper and lower subbands is relatively thick. As a result, the designs in this study feature a reduced oscillator strength, that is, the oscillator strength for 2.2 THz, 3 THz, and 3.8 THz lasers are 0.15, 0.2, and 0.24, respectively. To compensate

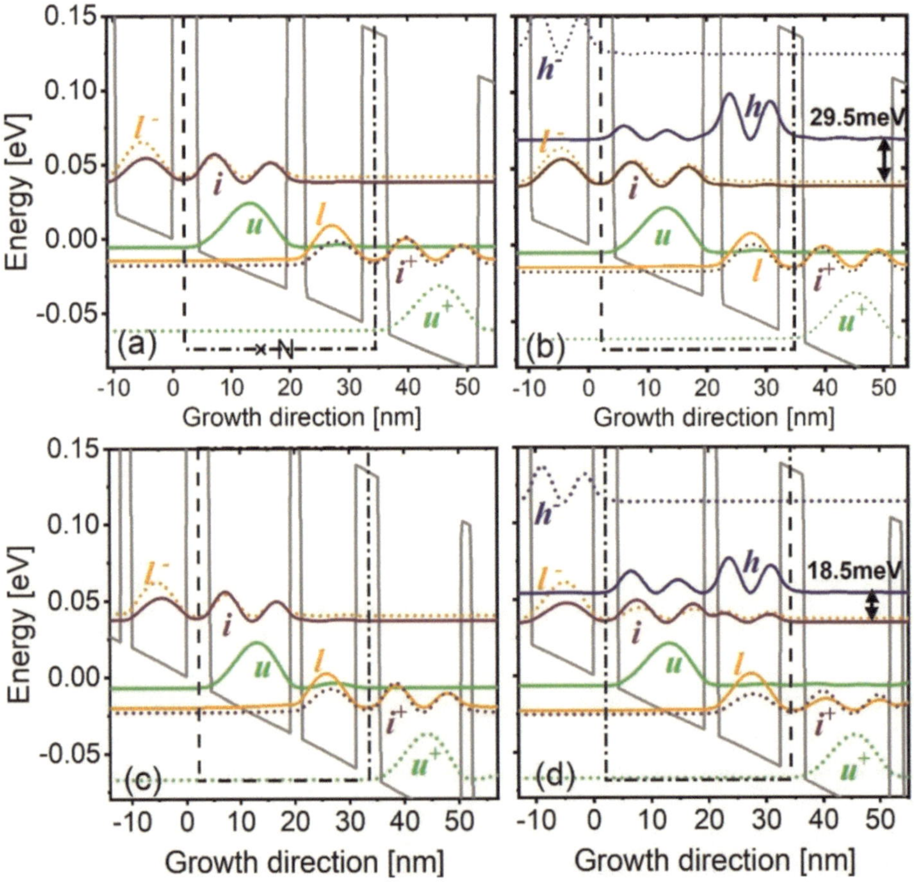

Figure 3.
Conduction band profile with the modulus square of the Wannier-Stark subbands under the operational bias, for the 2.2 THz design (a, b) and 3.8 THz design (c, d) including/excluding the high-lying subband h.

for such low oscillator strength, the doping level is correspondingly increased. Meanwhile, to avoid too strong Coulomb scattering effects, the periodic doping levels are balanced at 5.5×10^{10} cm^{-2}, 5×10^{10} cm^{-2}, and 4.7×10^{10} cm^{-2} (sheet doping density) for the 2.2 THz, 3 THz, and 3.8 THz designs, respectively.

Figure 4 shows the changes in population inversion ($\Delta n = n_u - n_l$) and the optical gain peak as functions of the lattice temperature, the plots are shown with the inclusion or exclusion of the high-lying subband h during modeling. 1) *Population inversion.* Regardless of the frequencies and lattice temperatures, the population inversion Δn increases when the subband h is included (**Figure 4-a1, b1** and **c1**). This differs from RT-QCLs in that the high-lying subbands are always treated as thermally activated electron leakage channels to reduce population inversion. In order to quantitatively estimate the changes, **Table 1** enumerates the magnitude of the changes in population inversion Δn. The table shows a 0.7% and 2.5% increasing share in Δn after including the subband h for the 2.2 THz and 3.8 THz designs at 300 K, respectively. The more increasing trend at 3.8THz can be explained by the quantum structures in **Figure 3**, that is, the upper well is positioned to encourage LO-phonon emission. Hence, the thickness of this well should be the largest. Likewise, the lower well should also be sufficiently wide to move down the lower laser subband l and satisfy the required THz radiation frequencies. Therefore, the injector subband i (the second excited subband in the upper well) and high-lying subband h (the second excited subband in the lower well)

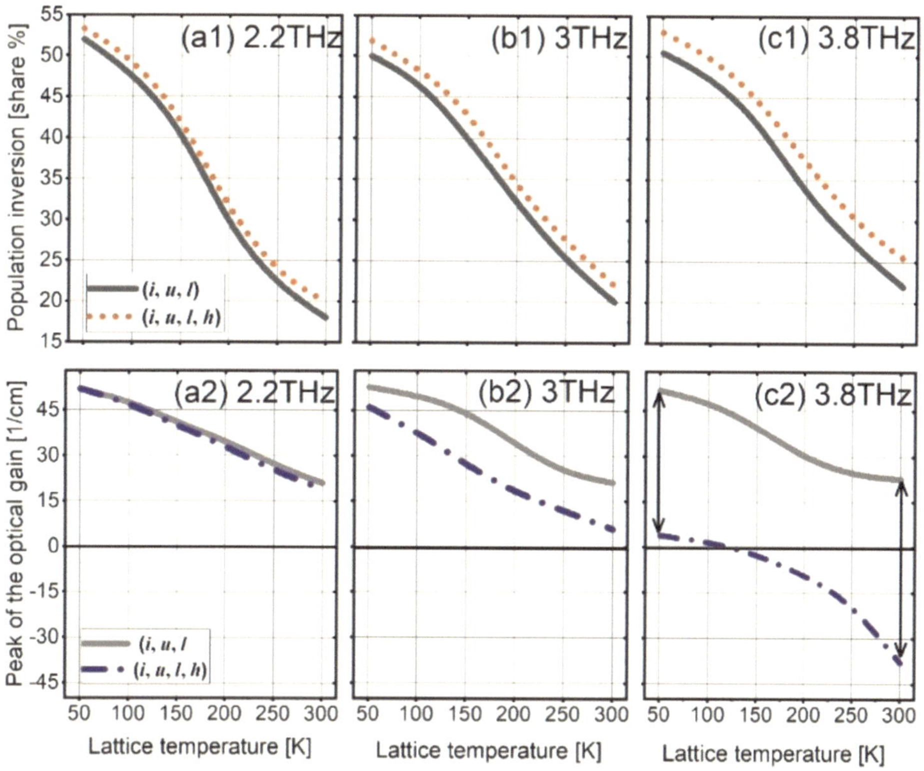

Figure 4.
Changes in population inversion (a1, b1, and c1) and real peak gain (a2, b2, and c2) in three different lasing frequencies SA-QCL designs (2.2 THz, 3 THz, and 3.8 THz). The x-axis corresponds to the lattice temperature, ranging from 50 K to 300 K.

Frequency (temperature)	Subbands number	n_h	n_i	n_u	n_l	Δn_{ul}	z_{ul} (nm)	G_p^* (cm^{-1})	G_p (cm^{-1})
2.2 THz (300 K)	(i, u, l)	—	14.5%	52%	33.6%	18.4%	2.4	21.5	21.5
	(i, u, l, h)	1.5%	13.6%	52.3%	32.6%	19.7% (0.7% ↑)	2.38	22.6 (1.1↑)	20.5 (1↓)
3.8 THz (300 K)	(i, u, l)	—	14.6%	54%	31.4%	22.6%	2.95	25	25
	(i, u, l, h)	2.6%	11.7%	55.4%	30.3%	25.1% (2.5%↑)	2.89	26.6 (1.6↑)	−38 (63↓)

Table 1.
Population shares at individual subbands (n_i, n_u, n_l, n_h) in the 2.2 THz and 3.8 THz designs, with the corresponding population inversions (Δn_{ul}). The total electrons population in each period is normalized as 100%. The dipole matrix elements for radiation transition between upper and lower laser subbands z_{ul} is also shown. A "normal" peak gain G_p^* is estimated based on the change in population inversion compared with the cases of (i, u, and l) and (i, u, l, and h), setting the value of (i, u, and l) as a standard. The peak gain G_p is the "real" value.

are close to each other. Especially for high lasing frequencies, the energy separation between those two subbands decreases more (29.5 meV for 2.2 THz and 18.5 meV for 3.8 THz in **Figure 2-b, d**). Meanwhile, the barrier between them cannot be excessively thick, as this barrier plays a role in tuning the oscillator strength between the laser subbands u and l. As a result, parasitic coupling is formed between i and l. The magnitude of this coupling can be quantified by the energy splitting between them ($2\hbar\Omega_{ih}$). The splitting energy is 3 meV and 10.5 meV in the 2.2 THz and 3.8 THz designs, respectively. Therefore, the high-lying subband h can act as an additional depopulation channel for the lower laser subband l in upstream period (noted that subbands i and l in neighboring periods are with full resonance alignment), this further depopulation can increase the population inversion. For higher lasing frequencies, this channel will be stronger owing to the enhanced coupling, leading to more increase in inversion share (i.e. 2.5% for the 3.8 THz designs). In addition, from **Table 1**, it can be observed that the non-equilibrium occupation of the high-lying subband h at 300 K is considerably low, that is, n_h = 1.5%/2.6% at 2.2 THz/3.8 THz. This demonstrates that the role of subband h is to act as a channel to partly redistribute the populations among the desired subbands, rather than itself storing a high share of population. 2) *Optical gain.* As shown in **Figure 4-a2, b2** and **c2**, after the inclusion of the high-lying subband h, the changes in the peaks of optical gain are inconsistent with the population inversion. In the 2.2 THz design, the peak gain is almost the same in the cases of (i, u, and l) and (i, u, l, and h), regardless of the temperature. However, when the lasing frequency is higher, the peak gain is strongly reduced and surprisingly even negative in the 3.8 THz designs, despite temperatures even as low as 115 K (**Figure 4-c2**). It should be noted that the inclusion or exclusion of the high-lying subband h, the dipole matrix elements z_{ul}, and the radiation transition linewidth Γ_{ul} remain almost unchanged. Therefore, following the semiclassical manner to predict optical gain, $G_p \sim \Delta n{*}z_{ul}^2/\Gamma_{ul}$, after the inclusion of high-lying subband h, an increased population inversion should improve the peak gain in the 3.8 THz design. **Table 1** presents the "nominal" gain peak G_p* for (i, u, l, and h) case at 300 K, which is estimated based on the changes in population inversion. In the 3.8 THz design, G_p* is 26.6 cm^{-1}, representing a 1.6 cm^{-1} increase over the (i, u, and l) case, whereas the "real" peak gain G_p is only −38 cm^{-1}, showing a sharp decrease of 63 cm^{-1} when compared with G_p*.

To study the inconsistency of the changes in population inversion Δn and the "real" optical gain G_p, additional data are extracted from the optical gain mappings and spectra. **Figure 5** shows the optical gain spectra for the 2.2 and 3.8 THz designs at both low/high temperatures (50 K/300 K). The black and colored solid curves represent the gain spectra for (i, u, and l) and (i, u, l, and h) cases, respectively. It is clear that the appearance of high-lying subbands h introduces strong parasitic absorption, as labeled by arrows a*/b* in both the 2.2 THz and 3.8 THz designs. The arrows a/b indicate the peak gain at the designed lasing frequencies. At both 50 K and 300 K, for the 2.2 THz design, the peak gain area is separated from this absorption, as a result, the net gain peak is not significantly different when comparing (i, u, and l) and (i, u, l, and h). By contrast, for the 3.8 THz design, this absorption can overlap with the peak gain area and induce a dramatic reduction in the net gain, regardless of the temperature. Considering the pairs of subbands, this absorption originates from the coupling between the injector subband i and high-lying subband h, where the energy separation is 29.5 meV in the 2.2 THz design and 18.5 meV in the 3.8 THz design. In particular, at high temperatures, electrons from the upper laser subband u (which shares most of the population in SA-QCL) will be thermally back to the injector subband i, thus enhancing this parasitic absorption. For the 3.8 THz design, as shown in **Figure 4-c2** labeled by the double-sided black arrows, the deviation of the peak gain between (i, u, and l) and (i, u, l, and h) is 45 cm^{-1} at 50 K, and 63 cm^{-1} at 300 K, respectively.

Figure 6 shows the gain mappings resolved based on the spatial position and lasing frequencies. Clearly, the emergence of parasitic absorption between the subbands i and h (**Figure 6b** and **d**) overlaps the gain in the 3.8 THz design (**Figure 6-d1** and **d2**).

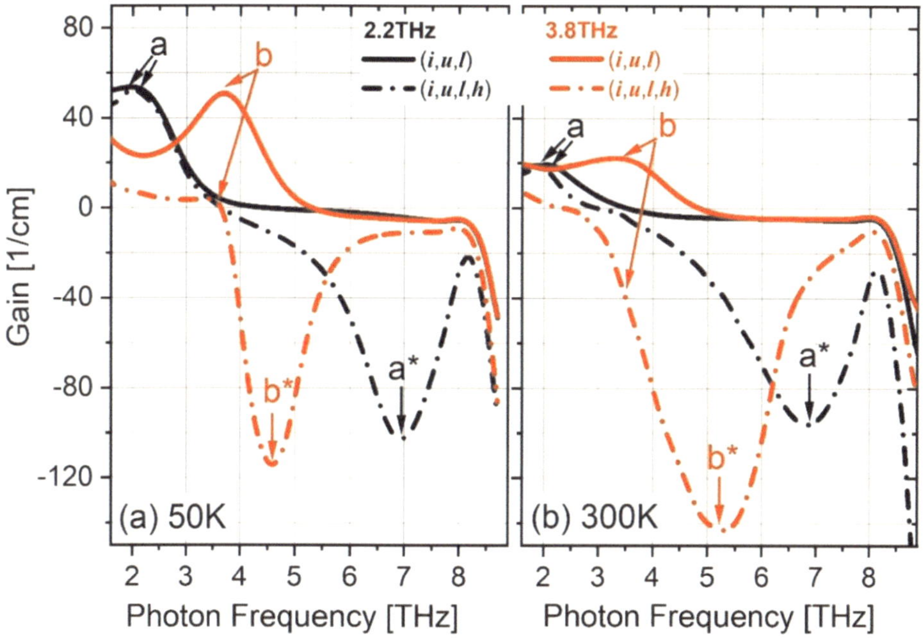

Figure 5.
Optical gain spectra of the 2.2 THz and 3.8 THz designs at 50 (a) and 300 K (b) under operational bias. Arrows a/b denote the peak gain, and arrows a/b* represent the parasitic absorption peak.*

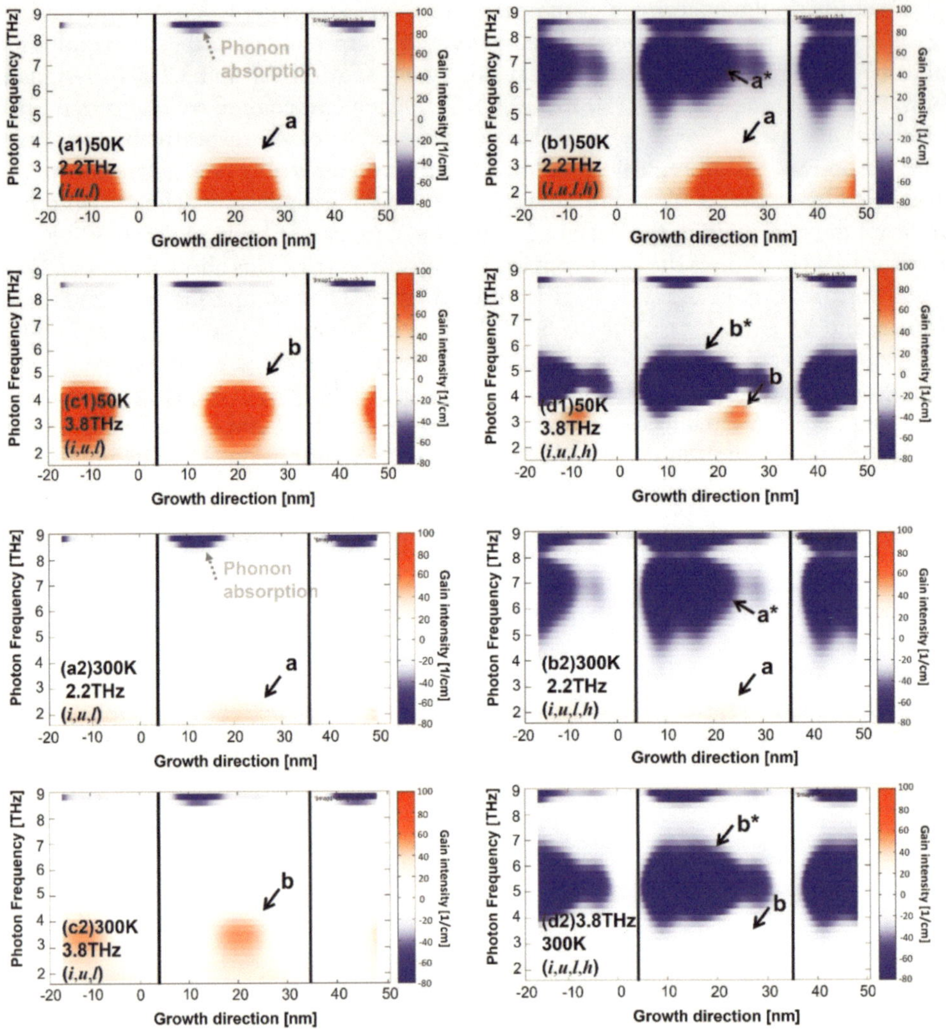

Figure 6.
Spatial and energy resolved gain mappings of the 2.2 THz design (a1, b1 at 50 K; a2, b2 at 300 K) and 3.8 THz design (c1, d1 at 50 K; c2, d2 at 300 K) under operational bias.

In general, 3–4 THz is the frequency band desired to achieve high-temperature operation [7, 25]. Therefore, this significant reduction in the optical gain reinforces the need for specific strategies to suppress this parasitic absorption, for example, by engineering the high-lying subband h.

Here, we study the feasibility of using step well to engineer the subband h. As shown in **Figure 7**, the use of AlGaAs in upper well (instead of GaAs) is proposed, and the Al composition in this ternary alloy can be controlled to tune the energy of parasitic absorption between subbands i and h. The upper well is set to make both the injector and upper laser subbands high in energy; meanwhile, the depopulation efficiency between them remains by keeping an energy separation same as the design in **Figure 3**. Consequently, the lower well can be narrowed correspondingly to satisfy the THz radiation frequencies. By doing this, the high-lying subband in the lower well is

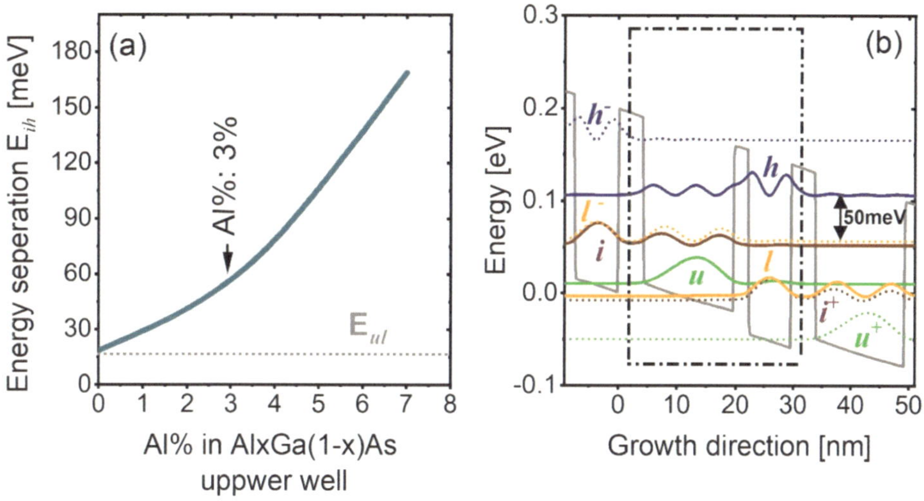

Figure 7.
Parasitic absorption energy between injector and high-lying subbands (i, h) as functions of the Al composition in upper well AlGaAs (a). Design structure (band profile and subbands) with Al% of 3% in upper well AlGaAs (b).

significantly upward. The parasitic absorption energy can be enlarged from 18.5 meV with AlGaAs (Al% = 0) to 170 meV with AlGaAs (Al% = 7%) (**Figure 7a**). A reprehensive design structure with Al% of 3% is shown in **Figure 7b**. Those strategies show possible pathway to guide the experiments, and the initial results [33] are convinced this strategy.

4. Conclusions

In summary, to estimate the optical gain in the SA-QCL design, the high-lying subbands need to be included, and the optical gain and absorption arising from any coupled pairs of subbands should be calculated, especially when the SA-QCL design is based on a simple quantum structure. Notably, as shown in this work, for SA-QCLs with two wells, population inversion increases when additional high-lying subbands are included, owing to the activation of more depopulation channels. However, an increase in the optical gain does not correspondingly occur if the lasing frequency exceeds 3 THz; instead, the peak gain undergoes a significant decrease, even below zero. The strong decoupling between population inversion and optical gain is ascribed to the emergence of parasitic absorption, which is caused by transitions between the desired subband and high-lying subband. Owing to the engineering limit permitted in the simple quantum structure, this parasitic absorption unavoidably overlaps with the optical gain, resulting in a reduction in the peak gain intensity. This overlap is more severe when the lasing frequency exceeds 3 THz. This finding reinforces the need for engineering the specific high-lying subbands to suppress the overlaps, thus realizing the two-well SA-QCL design experimentally. Here in the final part of this work, the feasibility by employing AlGaAs ternary alloys instead of GaAs for upper well is shown, and the small Al composition can dramatically enlarge the energy of parasitic absorption, thus almost removing the overlaps.

Author details

Li Wang
Research Center for Advanced Photonics, RIKEN, Sendai, Japan

*Address all correspondence to: li.wang@riken.jp

References

[1] Mittleman DM. Twenty years of terahertz imaging. Optics Express. 2018; **26**(8):9417-9431

[2] Scalari G, Walther C, Fischer M, Terazzi R, Beere H, Ritchie D, et al. THz and sub-THz quantum cascade lasers. Laser & Photonics Reviews. 2009;**3** (1–2):45-66

[3] Williams BS. Terahertz quantum-cascade lasers. Nature Photonics. 2007;**1**: 517-525

[4] Faist J, Capasso F, Sivco DL, Sirtori C, Hutchinson AL, Cho AY. Quantum cascade laser. Science. 1994;**264**(5158): 553-556

[5] Köhler R, Tredicucci A, Beltram F, Beere HE, Linfield EH, Davies AG, et al. Terahertz semiconductor-heterostructure laser. Nature. 2002;**417**:156-159

[6] Fathololoumi S, Dupont E, Chan CWI, Wasilewski ZR, Laframboise SR, Ban D, et al. Terahertz quantum cascade lasers operating up to $\sim$ 200 K with optimized oscillator strength and improved injection tunneling. Optics Express. 2012;**20**(4): 3866-3876

[7] Khalatpour A, Paulsen AK, Deimert C, Wasilewski ZR, Hu Q. High-power portable terahertz laser systems. Nature Photonics. 2021;**15**: 16-20

[8] Scalari G, Terazzi R, Giovannini M, Hoyler N, Faist J. Population inversion by resonant tunneling in quantum wells. Applied Physics Letters. 2007;**91**: 032103

[9] Kumar S, Hu Q. Coherence of resonant-tunneling transport in terahertz quantum-cascade lasers. Physical Review B. 2009;**80**:245316

[10] Dupont E, Fathololoumi S, Liu HC. Simplified density-matrix model applied to three-well terahertz quantum cascade lasers. Physical Review B. 2010;**81**: 205311

[11] Lee S-C, Wacker A. Nonequilibrium Green's function theory for transport and gain properties of quantum cascade structures. Physical Review B. 2002;**66**: 245314

[12] Kubis T, Yeh C, Vogl P, Benz A, Fasching G, Deutsch C. Theory of nonequilibrium quantum transport and energy dissipation in terahertz quantum cascade lasers. Physical Review B. 2009; **79**:195323

[13] Schmielau T, Pereira MF. Nonequilibrium many body theory for quantum transport in terahertz quantum cascade lasers. Applied Physics Letters. 2009;**95**:231111

[14] Callebaut H, Kumar S, Williams BS, Hu Q, Reno JL. Analysis of transport properties of terahertz quantum cascade lasers. Applied Physics Letters. 2003;**83**: 207-209

[15] Jirauschek C, Lugli P. Monte-Carlo-based spectral gain analysis for terahertz quantum cascade lasers. Journal of Applied Physics. 2009;**105**:123102

[16] Yasuda H, Kubis T, Vogl P, Sekine N, Hosako I, Hirakawa K. Nonequilibrium Green's function calculation for four-level scheme terahertz quantum cascade lasers. Applied Physics Letters. 2009;**94**:151109

[17] Yamanishi M, Fujita K, Edamura T, Kan H. Indirect pump scheme for quantum cascade lasers: Dynamics of electron-transport and very high T_0-

values. Optics Express. 2008;**16**(25): 20748-20758

[18] Kubis T, Mehrotra SR. Design concepts of terahertz quantum cascade lasers: Proposal for terahertz laser efficiency improvemencts. Applied Physics Letters. 2010;**97**:261106

[19] Kumar S, Chan CWI, Hu Q, Reno JL. A 1.8-THz quantum cascade laser operating significantly above the temperature of $\hbar\omega$/kB. Nature Physics. 2011;7:166-171

[20] Dupont E, Fathololoumi S, Wasilewski ZR, Aers G, Laframboise SR, Lindskog M, et al. A phonon scattering assisted injection and extraction based terahertz quantum cascade laser. Journal of Applied Physics. 2012;**111**:073111

[21] Scamarcio G, Troccoli M, Capasso F, Hutchinson AL, Sivco DL, Cho AY. High peak power (2.2 W) superlattice quantum cascade laser. Electronics Letters. 2001;**37**(5):295-296

[22] Liu T, Kubis T, Wang QJ, Klimeck G. Design of three-well indirect pumping terahertz quantum cascade lasers for high optical gain based on nonequilibrium Green's function analysis. Applied Physics Letters. 2012; **100**:122110

[23] Scalari G, Hoyler N, Giovannini M, Faist J. Terahertz bound-to-continuum quantum-cascade lasers based on optical-phonon scattering extraction. Applied Physics Letters. 2005;**86**:181101

[24] Amanti MI, Scalari G, Terazzi R, Fischer M, Beck M, Faist J, et al. Bound-to-continuum terahertz quantum cascade laser with a single-quantum-well phonon extraction/injection stage. New Journal of Physics. 2009;**11**:125022

[25] Bosco L, Franckié M, Scalari G, Beck M, Wacker A, Faist J. Thermoelectrically cooled THz quantum Cascade laser operating up to 210 K. Applied Physics Letters. 2019;**115**:010601

[26] Kubis T, Vogl P. Assessment of approximations in nonequilibrium Green's function theory. Physical Review B. 2011;**83**:195304

[27] Lake R, Klimeck G, Bowen RC, Jovanovic D. Single and multiband modeling of quantum electron transport through layered semiconductor devices. Journal of Applied Physics. 1997;**81**: 7845-7869

[28] Klimeck G, Lake R, Blanks DK. Role of interface roughness scattering in self-consistent resonant-tunneling-diode simulations. Physical Review B. 1998;**58**: 7279-7285

[29] Lake R, Klimeck G, Bowen RC, Fernando C, Moise T, Kao YC, et al. Interface roughness, polar optical phonons, and the valley current of a resonant tunneling diode. Superlattices and Microstructures. 1996;**20**(3):279-285

[30] Grange T. Electron transport in quantum wire superlattices. Physical Review B. 2014;**89**:165310

[31] Grange T. Contrasting influence of charged impurities on transport and gain in terahertz quantum cascade lasers. Physical Review B. 2015;**92**:241306(R)

[32] Grange T. Nanowire terahertz quantum cascade lasers. Applied Physics Letters. 2014;**105**:141105

[33] Wang L, Lin TT, Hirayama H. Two-well indirect pumping terahertz quantum cascade lasers. Applied Physics Express. (to be submitted)

Chapter 2

Fundamentals of Femtosecond Laser and Its Application in Ophthalmology

Sana Niazi and Farideh Doroodgar

Abstract

Modern advancement in lithographic technology, injection molding, and nano-imprinting has improved the patterning of small structures, resolution, productivity, and materials. Ultrafast laser micro/nano-manufacturing technologies, including nano- and femtosecond lasers, have the advantage of high precision as a result of suppressed heat diffusion to the surroundings. This precision imposes strict requirements on the temporal characteristics of laser pulses. Ultrafast lasers also have advantages in terms of technique, application, and processing. Femtosecond laser (FSL) uses photo disruption to form micro-cavitation bubbles within the cutting plane. The controllable spatiotemporal properties of FSL make it applicable for the three-dimensional fabrication of transparent materials. Using smart materials to create 3D microactuators and microrobots is a newfound application of FSL processing, which enables the integration of optical devices with other components and is practiced in new applications, such as 3D microfluidic, optofluidic, and electro-optic devices. We discuss mechanisms and methods of FSL (including digital micromirror devices, different processes, and interferences). Microlens arrays, micro/nanocrystals, photonic crystals, and optical fibers all have applications in the production of optical devices. Using FSLs, one may create scalable metamaterials with multiscale diameters from tens of nanometers to centimeters. The huge potential of FSL processing in various fields, such as machinery, electronics, biosensors and biomotors, physics, and chemistry, requires more research.

Keywords: lasers, lasers, solid-state, corneal surgery, laser, ophthalmology, cornea, optics and photonics

1. Introduction

Ophthalmology has faced substantial changes by the advances in medical technology. These advances include developments in diagnostic and therapeutic strategies. One of the important advances of recent years, which has changed the field of ophthalmology, is the laser. After the development of the early solid-state lasers, FSL was developed as a near-infrared laser with ultra-short pulse duration, which enabled safety, precision, and predictability of ophthalmic surgeries and improved the postoperative outcome of patients. It has also been used for the fabrication of several optic devices, which helped to improve the precision of diagnosis and treatment in the field of ophthalmology. The ongoing advances in FSL technology since its introduction are making the newer machines with better quality, stability, and fewer complications. Therefore, it is suggested that researchers and clinicians keep themselves updated about the advances in FSL technology and make use of the updated technologies to benefit from its advantages.

Nano/micro-devices have diverse applications in optics, nanophotonics, Opto-electronics, and biomedical engineering. As the transparent layers of the eye do not absorb electromagnetic radiation, light does not alter them, but the light energy is absorbed by them, causing focal tissue disruption. Laser systems with different wavelengths have been developed based on these principles with several applications in ophthalmology. The femtosecond laser (FSL) is a near-infrared laser with a wavelength of 1053 nm and a diameter of 0.001 mm. The term "femtosecond" refers to the pulse duration of one billionth of a second, fired in the nanosecond domain. This ultra-short pulse duration allows for smaller shock waves, focus to a 3-μm spot, no heat development, and less damage to the adjacent and superficial tissue (collateral damage) [1].

FSL works by the production of a different tissue interaction, known as photodisruption. Within the cutting plane, micro-cavitation bubbles are formed discretely. Non-linear absorption of laser energy is obtained by using many photons at the same place and time. This multiphoton effect causes the tissue to absorb energy as high as it results in optical breakdown. This process of photodisruption creates plasma, acoustic shockwave, thermal energy, and cavitation bubble, which expands at supersonic speed, slows down, and then implodes and eventually forms a gas bubble composed of carbon dioxide, water, nitrogen, and other elements. This will break the limitations of traditional fabrication methods, which results in two–threefold better precision than cell surgery using continuous-wave irradiation [2]. In addition, the quality of fabrication is also improved in FSL, due to the non-equilibrium and non-thermal absorption phase transitions, resulting in reduced heat-affected zones, cracks, and recast layers. Furthermore, FSL requires no mask, vacuum, or reactive gas environment. These characteristics make FSL applicable in different manufacturing processes, such as telecommunications technology, biotechnology, pharmaceuticals, aerospace, and environmental industries. The non-linear and material-independent absorption of femtosecond lasers makes them ideal for creating complex 3D structures in composite substrates with nanometer-scale precision [3].

Modern laser technology is rapidly improving in terms of resolution, productivity, and materials. Using smart materials to create three-dimensional (3D) microactuators and microrobots is a newfound application. With femtosecond laser processing, optical devices can be seamlessly integrated with other components, enabling new applications such as 3D microfluidic, optofluidic, and electro-optic devices.

We discuss mechanisms and methods for femtosecond laser, including digital micromirror devices, different processes, and interferences. Microlens arrays, micro/nanocrystals, photonic crystals, and optical fibers all have applications in the production of optical devices. Using femtosecond lasers, one may create scalable metamaterials with multiscale diameters from tens of nanometers to centimeters. Improved methods may result from the availability of effective femtosecond laser surgical equipment for any refractive surgery processes, Photo disruptive ultrasonic lens surgery as a combo approach in any medium to altering the power of the intraocular lens (IOL) postoperatively. The huge potential of femtosecond laser processing in fields such as machinery, electronics, biosensors and biomotors, physics, and chemistry requires more research.

2. Historical background for development of femtosecond laser systems in ophthalmology

Various laser sources were developed and pioneered by solid-state and organic dye lasers in the 1960s and early 1970s. In 1979, short-pulse lasers at near-infrared wavelengths were used in ophthalmology for the treatment of posterior capsule opacification after surgery (by Aron-Rosa). In the 1980s, research continued on non-linear effects in optical fibers and near-infrared spectral region; in 1982, the first Titane (Ti): sapphire laser was built by Moutan, which had a wide tuning range (680–1130 nm), which was based the tunable FSL source. Following the use of the coupled-cavity mode-locking technique with a Ti: sapphire laser to constitute the generation of sub-100 fs pulses, a new generation of FSL was generated, which matched the cavity containing the 'non-linear element' with interferometric precision to the master cavity of the color-center laser oscillator. With peak optical pulse powers of >5 MW from a Ti: sapphire laser, compared with the dye-laser pulse peak powers, oscillator-amplifier combinations were no longer required, but for using the optical Kerr effect, cavity-design parameters were required (**Figure 1**). In 1984, FS dye laser was used in oscillator-amplifier configurations, and in 1989, Stern and colleagues found that reducing the pulse duration of near-infrared laser from the nanosecond to the picosecond (10^{-12}) and then FS (10^{-15}) resulted in higher precision ablation profiles and less collateral damage. FS belongs to ultra-fast or ultra-short pulse lasers with a beam diameter of <8 microns in the near-infrared spectral region and is capable of producing smaller shock waves and cavitation bubbles that affect a tissue volume about 10^3 times less than picosecond-duration pulses [1].

The first prototype of the ophthalmic surgical FSL system was developed by Dr. Jujasz and colleagues and was clinically used by Dr. Kurtz at the University of Michigan in the early 1990s. The IntraLase Pulsion FSL was approved by the U.S. Food and Drug Administration (FDA) for lamellar corneal surgery in Jan. 2000, and the first commercial laser was introduced to the market in 2001 for the creation of the corneal flap in

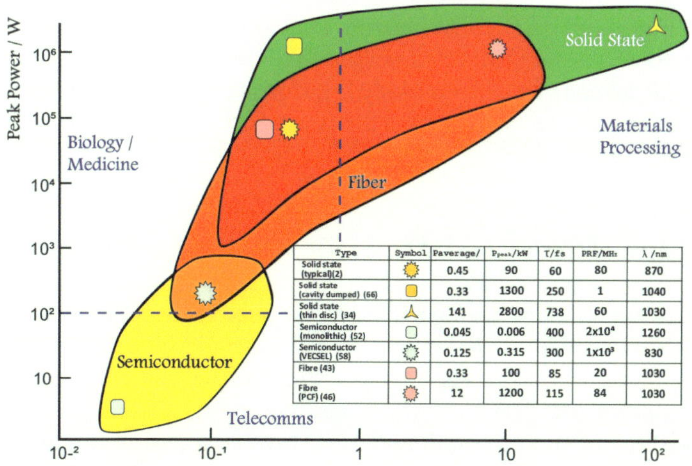

Type	Symbol	Paverage/	Ppeak/kW	T/fs	PRF/MHz	λ /nm
Solid state (typical)(2)		0.45	90	60	80	870
Solid state (cavity dumped) (66)		0.33	1300	250	1	1040
Solid state (thin disc) (34)		141	2800	738	60	1030
Semiconductor (monolithic) (52)		0.045	0.006	400	$2x10^4$	1260
Semiconductor (VECSEL) (58)		0.125	0.315	300	$1x10^3$	830
Fibre (43)		0.33	100	85	20	1030
Fibre (PCF) (46)		12	1200	115	84	1030

Figure 1.
The diversity of ultrashort pulse laser oscillators.

refractive surgery, i.e. laser in situ keratomileuses (LASIK). The advantages of this system caused FSL to replace the mechanical cutting devices shortly. In 2002, Advanced Medical Optics (IntraLase FS, Irvine, California, USA) fired a 10–kHz laser. In 2007, Ziemer FEMTO LDV™ introduced new low pulse energy with high frequency. The current IntraLase system has a pulse rate of 60 kHz, which enables shorter flap-cutting times, less energy to cut the flap, and closer separation of the spots and lines. The fifth generation of the IntraLase FS system fires at 150 kHz with high-precision computer control of the parameters, which enables cutting flaps in less than 10 seconds with a variety of geometric shapes, depths, diameters, wound configurations, energy, spot sizes, and spot separation, allowing for precise corneal cutting.

FSL has also been used for cataract surgery since 2008; the LensX™ system, approved for this procedure by FDA in 2009, opened another sector of ophthalmic FS-laser application. The early version operated at 33 kHz pulse frequency and 6–15 μJ energies. LensX was then integrated with Alcon, and similar products were launched by multiple manufacturers. In 2014, the first low pulse energy FSL system was introduced for cataract and cornea surgery, the Ziemer FEMTO LDV Z8™ [4].

Simultaneous with the development of FSL, optical coherence tomography (OCT) was also described (1988), which provided non-invasive 3D in vivo imaging with fine resolution (microscopic resolution of 5–20 μm) in both lateral and axial dimensions at a micrometer level. Several variations have been developed for OCT since its introduction and are currently used with FSL systems in most modern cataract procedures after docking the laser interface to the eye.

Fourier Domain Optical Coherence Tomography (FD-OCT) employed a fixed reference arm length but a spectrometer with a linear detector array instead of a single detector. In this scenario, optical path length variations between interferometer arms cause periodic interference modulation. By Fourier transformation, the measured spectrum can yield an A-scan. "Sweep-source" (SS) OCT is an improved frequency-domain OCT variant. A tunable light source with a "sawtooth" frequency profile over time is used with a fast single-pixel detector instead of a spectrometer. After docking the laser interface to the eye, most current cataract fs-lasers do three-dimensional OCT scans.

Also, a 3D confocal structured illumination is used with the Scheimpflug camera, which was first described by Theodor Scheimpflug in 1904, and is currently used in LENSAR™ system. Therefore, the integration of FSL processing with optical devices enabled new applications in ophthalmology [5]. Considering the day-to-day advancements of FSL, in this chapter, we describe the latest micromachining advancements in optical systems and devices.

3. Characteristics of femtosecond laser

Before we describe the processing methods, we need to know about the characteristics of FSL. Compared with the conventional laser microfabrication techniques, which use continuous-wave lasers or long-pulse lasers, the processing of ultrafast lasers, including nano- and femtosecond lasers, has several advantages. The short laser pulse (< electron–phonon coupling time of the laser-matter interactions) makes the laser energy absorbed by electrons and rapidly transferred, which reduces the thermal effect of ultrafast laser. This characteristic of ultrafast lasers is one of the fundamental advantages, which results in higher precision that enables the

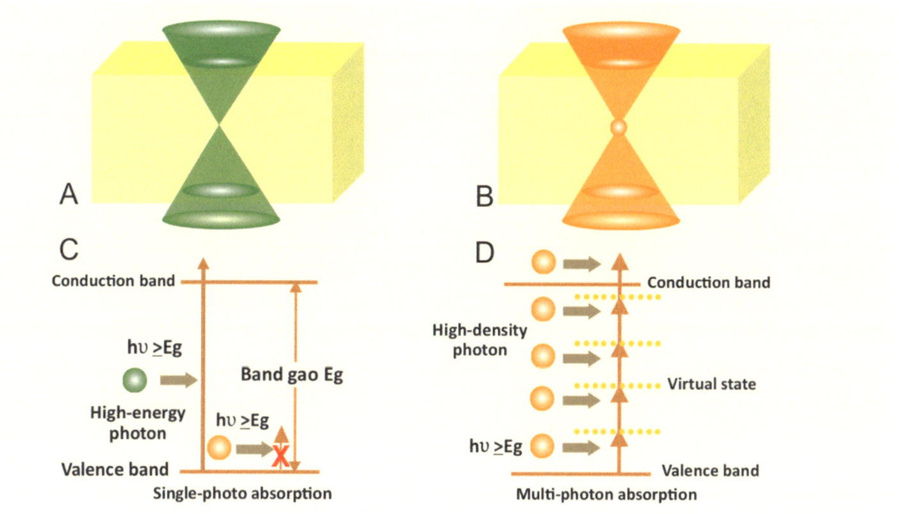

Figure 2.
Laser processing of a transparent material by single and multiphoton absorption and their electron excitation process.

fabrication of fine structures. This characteristic is especially important when the peak intensity of the laser has to be controlled near the ablation threshold with a low repetition rate.

The second important advantage of ultrafast lasers is related to the non-linear electron excitation mechanism that is unique to these lasers and results in the induction of strong absorption, even in transparent materials. In the conventional linear single-photon, absorption (long-pulsed or CW lasers) requires photon energies greater than the material's band gap for excitation of an electron from the valence band to the conduction band by absorbing a single photon. Because of the surface absorption, the transparent materials cannot be internally modified. In ultrafast lasers, absorption of the extremely high density of photons (multiphoton) enables the interaction between the laser pulse and a transparent material only near the focal point (**Figure 2**). Therefore, no out-of-focus absorption occurs by focusing FSL beam on transparent bulk material. In the next subtitle, we describe the mechanism of non-linear and multiphoton absorption processes of FSL in detail.

The combination of the two above-mentioned characteristics (heat-affected zones suppression and non-linear multiphoton absorption) shapes the third advantage of ultrafast laser processing, i.e. a resolution far beyond the diffraction limit. This factor is related to the Gaussian spatial profile of the laser beam intensity, which makes the effective absorption coefficient for n-photon absorption proportional to the nth power of the laser intensity that causes a narrower absorbed energy distribution for multiphoton absorption. The fabrication resolution can be additionally improved by adjusting the laser intensity at a threshold above which a reaction occurs on absorption.

The last important characteristic of FSL to be mentioned here is the spatially selective manner of tuning or altering the physical and chemical properties of a material. This characteristic, along with the multiphoton effect, is used in 3D femtosecond laser direct writing for the integration of multiple functions on a single substrate [6].

4. Mechanisms of femtosecond processing

Understanding the physical mechanisms and processes of FSL is important, as they differ between FSL fabrication (including phase change and material removal) and traditional manufacturing methods, which are essentially determined by laser–electron interactions. Hence, the regulation of laser-electron interactions or electrons dynamics is critical to the future development of femtosecond laser manufacturing, which poses a challenge for measuring and controlling at the electron level during fabrication processes. Hence, the development of theory and observation systems must be synchronized with the development of laser fabrication methods and applications. In ophthalmic surgery, laser energy is channeled as efficiently as possible. The flap thickness is determined by placing a sterile plastic foil between the laser and the cornea. The computer keeps all corneal cuts suctioned up in a total vacuum time of <40 seconds. Laser cuts the tissue by two mechanisms; some laser pulses vaporize small amounts of tissue by photodisruption process. Vaporized tissue causes multiple intrastromal cavitation bubbles of microplasma, composed of water and carbon dioxide. The bubbles disrupt the tissue at a larger radius than the plasma created at the laser focus, which dissociates the tissue and creates a lamellar corneal dissection plane. Other lasers create a dissection plane using the desired pattern (e.g., a raster or spiral pattern), controlled by the surgeon using laser software. The nature of the cutting processes differs based on the laser-tissue interaction parameters, which include (1) Pulse energy, (2) Pulse repetition rate, (3) Pulse duration, (4) Wavelength, (5) Focusing power, (6) Focus spot shape, and (7) Spatial pulse spacing.

The laser energy used may be a high or low pulse. Lasers with high pulse energy and low frequency (with pulse energies in J and repetition rates in kHz) were used earlier (**Figure 3**). Low repetition rate and pulse overlap reduce tissue bridging. In the high pulse energy laser group, the mechanical forces drive the cutting process by the expanding bubbles, which disrupt the tissue at a larger radius than the plasma created at the laser focus. Modern FSLs use low pulse energy by shortening the pulse duration

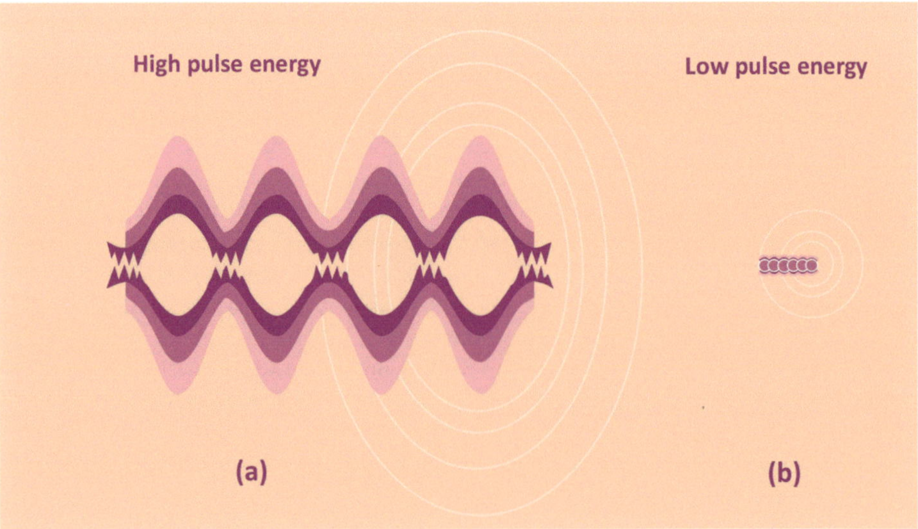

Figure 3.
High pulse low-frequency energy.

or reducing the focal spot size to reduce the side effects at a given wavelength. In the low pulse energy group, high pulse frequencies are applied (MHz range), which helps achieve a cutting speed as effective as in the high energy laser group. Tissue evaporation inside the plasma effectively separates tissue without a need for secondary mechanical tearing effects [5].

Several parameters play a role in the interaction between the laser and the material, including chemical, thermal, and mechanical effects via free-electron generation. Ablation of material is caused by thermal damage resulting from impurities and defects in the samples. FSL intensities ≥1014 W/cm^2 are anticipated to have the same ablation mechanisms for metals and dielectric materials. Gamaly et al. also verified the ablation threshold value and ablation velocity formula of the metal and dielectric material. This interaction also differs between metallic and non-metallic materials, as metallic materials have a large number of free electrons while non-metallic materials do not. The absorbed FSL by free photons of metallic materials results in heated photons and their collision with other electrons that transfer energy to each other and increase the interaction with the lattice. The heated lattice results in the phase transition of the material, at picosecond (10^{-12} s) to tens of nanoseconds (10^{-8} s), resulting in micro- to millisecond-scale plasma. Conversely, in non-metallic materials, electrons are bound in the valance band. The processing of FSL in non-metallic materials includes ionization and phase change. Ionization involves photoionization (which includes tunnel ionization and multiphoton ionization) and impacts ionization mechanisms. Photoionization is the main mechanism of seed-electron generation. The super-strong electromagnetic field generated by FSL reduces the Coulomb potential barrier of valence-band electrons by the tunneling effect, which results in the transfer of electrons into the conduction band and become free electrons. Multiphoton ionization refers to the absorption of multiple photon energies by the valence-band electrons, which obtain higher levels of kinetic energy, collide with other valence electrons, and transfer into the conduction band, causing a chain reaction similar to avalanche, resulting in free electrons. In other words, avalanche ionization involves free carrier absorption followed by impact ionization. As the electron's energy exceeds the conduction band minimum (more than the band-gap energy), it can ionize another electron from the valence band, which results in two excited electrons that can be heated by the laser field through free carrier absorption. This process repeats as long as the laser field is present and intense enough, leading to an electronic avalanche.

The phase change follows the first part of ionization and electron heating. In this part, the accumulation of a large number of free electrons inside the non-metallic material and the lattice-electron interaction results in the exhibition of transient metallic characteristics. In the temporal scale of picosecond (10^{-12} s) to tens of nanoseconds (10^{-8} s) and the spatial scale of tens-of-nanometers-to-tens-of-micrometers, the phase change includes thermal phase transitions (melting and gasification; when the lattice temperature rises above the melting point of the material) and non-thermal phase transitions (Coulomb explosions and electrostatic ablation; based on plasma expansion). The pump-probe technique can be used to observe the ablation processes on this temporal and spatial scale. The thermal damage depends on the pulse width and intrinsic parameters of the material (like melting point, thermal expansion coefficient, thermal conductivity coefficient, and tensile strength). As a general rule, only the fraction of energy within a laser pulse, absorbed inside the tissue, is responsible for interactions with tissue. The emitted laser energy is redistributed from a surgical laser device at the end of the tissue dissection process (**Figure 4**), and only non-linear

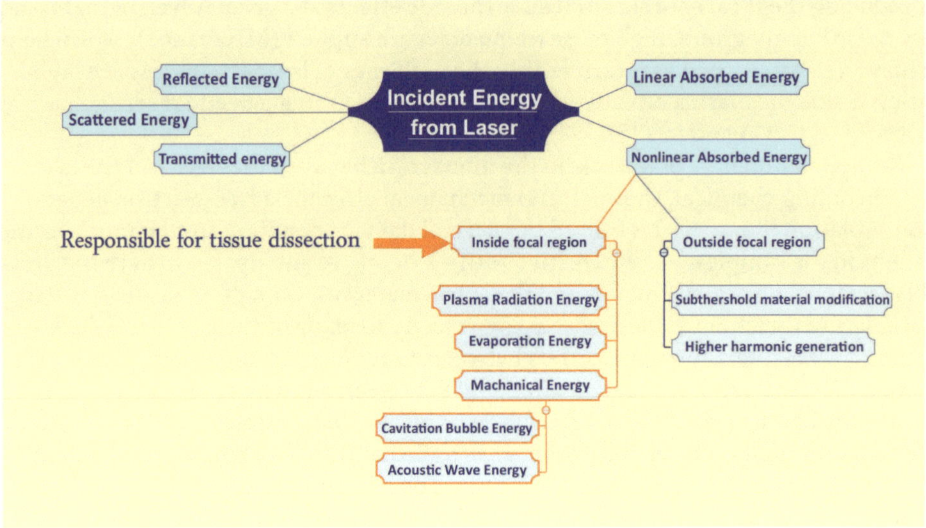

Figure 4.
Redistribution of energy in a pulsed laser process for tissue dissection.

absorbed energy, which constitutes only 10–15% of FS pulses, contributes to the tissue dissection process. Similarly, in transparent dielectric materials, the laser creates free carriers inside materials by non-linear absorption processes. For transparent materials, similar models have been described (as for non-metallic materials), including multi-photon, tunnel ionization, and avalanche.

It has to be noted that the material-laser interaction is complex, and different theoretical models have been suggested for its description, including time-dependent density function theory, molecular dynamics model, plasma model, and improved two-temperature model. But, each of these has its own limitations in special and temporal scales. For example, molecular dynamics combines the thermal and non-thermal phase transition mechanisms to explain material ablation due to lattice phase transition. However, some aspects have been described simply in this model, such as the alteration of the interaction force between atoms during the ablation process and approximate simplifications in the first-principles model, such as a time-dependent exchange-correlation, which influence the accuracy of theoretical predictions. The plasma model is used to describe photon absorption, plasma generation, and recombination of electronic systems before lattice phase transitions and offers a good explanation for the non-metallic ionization process. However, the phase transition process is not discussed in this model. Two-temperature model, described by Anisimov and colleagues for the interaction between ultrafast laser pulses and solids, is widely used for the prediction of the electron and phonon temperature distributions in laser processing. However, in this model, the electron density is set to a constant value; therefore, it is only applicable in metallic materials [3].

The models were extended by other researchers for more accurate calculations, such as the multiscale theoretical model and observation system, applicable for problems associated with each temporal scale to cover the overall scope of laser fabrication. The imaging system, such as sequentially time all-optical mapping photography, can be integrated with FSL fabrication system to realize real-time continuous observation and feedback of the fabrication process. Other theoretical models, based on simplified rate equations or a kinetic approach to Boltzmann's equations, have also been described. Models considered for the multiphoton non-linear optical process include two-photon absorption (which occurs when two photons are absorbed simultaneously by material) and two-photon polymerization (TPP; which occurs when two photons are absorbed on photosensitive material). This process changes the material and leads to polymerization by activating the photoinitiator-activated free radicals in the resist. Fabrication resolution of true 3D micro/nanostructures by TPP varies based on the laser energy, exposure duration, and concentration of the free radical [7].

Micro/nanostructures induced in transparent materials have been classified by Qiu et al. into four types based on optical coloration, refractive index modification, micro-hole creation, and micro-crack creation. FS pump-probe interferometry technique, with 100 fs temporal resolution, allows measuring the modification of refractive index induced by ultra-short intense laser pulses. When a dielectric material is subject to intense FSL, a large number of excited electrons may be generated by laser pulses, which produce intrinsic defects that make FSL an ideal tool for high-precision material processing. A complex secondary process of high-temperature and high-pressure plasma at the tightly focused laser point would induce a phase or structural modification of the material. The splitting and self-(de)focusing of FSL pulses inside dielectrics are the topics that researchers have used for the calculation of the excitation and relaxation channels of FS-induced carriers. Despite various models and explanations for laser-material interactions, the dynamics of specific phenomena are realized only partially. To quantify dynamics of laser-excited carriers, direct visualization of FSL dynamics using ultrafast imaging and spectral interferometry techniques has been designed and implemented successfully [8].

5. Systems and methods of femtosecond laser fabrication

Two important types of FSL micromachining include direct writing, parallel method, laser pulse shaping in the temporal and spatial domain, laser frequency modulation, and laser-pulse-coordinated shaping in temporal/spatial/frequency domain based on electrons dynamics control. The direct writing processing method, the commonly used basic processing method, is a serial fabrication process, and the increased numerical aperture results in achieving augmented resolution. The high resolution, excellent flexibility, and quality make this method suitable for fabricating 3D microstructures within transparent materials, while in the parallel method, a slit-oriented parallel is placed in the scanning direction before the objective lens, and the slit's orientation should be adjusted along with the scanning direction; therefore, the aspect ratio of the hollow microchannel and the laser efficiency is low. This method can only fabricate periodic structures. Another difference between the two methods is the throughputs; parallel microprocessing can realize high throughputs and is appropriate for large-scale FSL micromachining. But direct writing method has the limitation of low throughputs, which has been improved by the use of high-average-power and high-repetition-rate FSLs in recent years [3, 6, 9].

The choice of method to be used depends on what is required in that particular application. For each specific application, we have to consider the relation between working distance and fabrication resolution, as well. The fabrication resolution depends on their numerical aperture (NA). The focal spot diameter is inversely proportional to the NA (in the lateral direction, 1.22 λ/NA). Consequently, we require an objective lens with NA of ~0.5 for 3D machining. Then, a high-NA objective lens (e.g., oil immersion objectives), which usually has a short working distance, is applicable for surface nanostructuring or TPP (which involves surfaces and thin samples); while they are not appropriate for the fabrication of 3D microstructures deep in a substrate [6].

If the spatial and temporal aspects of laser pulses are simultaneously focused, the laser components will be separated spatially in space before they enter the objective lens and then overlap at the spatial focal point focus of the objective lens. For processing a symmetric spherical light intensity distribution and improved axial resolution, the shortest pulse duration should be confined to the spatial focus; this approach increases the resolution. In contrast, for application in microfluidics, we require an improved aspect ratio of a hollow microchannel, achieved by combining temporally- and spatially-shaped laser pulses to enhance the etching depth of the microchannel by a factor of 13, enabling high-throughput fabrication of ultra-high-aspect-ratio hollow microchannel [9]. In the following, we explain the FSL processing systems in detail.

6. Direct writing

A typical FSL direct writing system is composed of a laser source, a beam control/shaping system, a microscope objective, or an aspherical lens (**Figure 5**). A tightly focused spot and a high-precision XYZ translation stage are produced by the lens (which determines the fabrication resolution) and controlled by a computer for 3D translation of the sample. As FSL pulses possess broad spectra, objective microscope lenses are frequently used for minimizing both spherical and chromatic aberrations. For controlling the repetition rate, an electro-optic or acoustic-optic device is employed in the beam control/shaping system, which can also be used for creating burst mode, a spatial or temporal pulse shaper, a tunable attenuator, and a mechanical shutter. For real-time monitoring of the fabrication process, a charge-coupled device camera connected to a computer can be installed above the focusing lens. A tunable

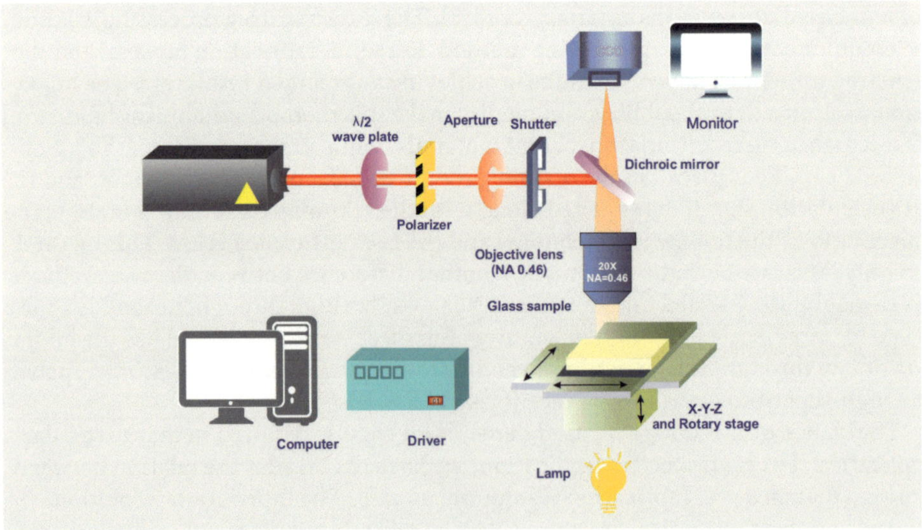

Figure 5.
Schematic image of femtosecond laser direct writing system.

attenuator is another component of the direct writing system, which controls the power and ensures stable operation of laser parameters such as the pulse width, pulse energy, and pointing direction. Autocorrelator devices have also been developed for characterizing the spatiotemporal profiles, including temporal duration or structure of ultrashort pulses [6].

Direct writing has two methods, depending on scanning directions (geometries): horizontal (where the sample is moved perpendicularly to the laser beam, commonly used to fabricate surface structures) and vertical (where the sample is moved along the direction of laser beam irradiation; it can be performed from the upper or lower surface). The greater longitudinal depth of the focal spot intensity distribution than its transverse dimension results in an asymmetric cross-section of the laser modification zone. An objective lens with a larger NA can reduce this difference, resulting in tight focusing and adjust the spatial distribution of pulse intensity. When direct longitudinal writing is processed from top to bottom, the laser beam is affected by the scattering of ablated materials, which reduces fabrication quality [3].

Horizontal direct writing is appropriate for the fabrication of ultra-slip surface porous network structures with excellent performance in liquid repellency and cell proliferation resistance, hierarchical structures, nano grooves, nanoholes, and 3D resonant optical cavities, widely used in applications such as hydrophilic and hydrophobic treatments, optical communication and sensors and biomedicine. Direct writing is also applicable for the fabrication of two-dimensional (2D) and 3D microstructures, either through moving the 3D transform stage (applicable to array machining and applications not requiring high precision) or through a galvanometer combined with the transform stage. In the former method, even by using Bessel Beam, which improves efficiency, rapid and flexible micromachining of complex microstructures cannot be achieved at a large scale. On the contrary, the latter method, scanning galvanometers and piezo has highly-developed high power, high repetition frequency, and miniaturized FSL, which makes it appropriate for high throughput and high-resolution micromachining both axially and laterally; this method is beneficial for the commercialization of FSL micromachining. TPP technique can be used for the construction of complex mesoscale 3D microstructures with nanoscale precision, like entire hollow devices and spiral phase plates. Direct writing also has another great potential, such as large-scale multifunctional smart materials with 3D gradient densities that can be widely employed in the processing of four-dimensional (4D) smart sensors and reconfigurable micromachines, actuators, and soft robots applicable in biomedicines (which use hydrogels; soft materials with high biocompatibility and deformability).

Besides the advantages of FSL direct writing, it also has some limitations, which makes it inappropriate for oil-working conditions used for biomaterials (e.g., cells and tissues). The serial process nature of this method also results in low throughput, although we can increase the scanning speed by replacing the XYZ motion stage with a galvo scanner. The recent ultrafast laser systems have high power and high repetition rate and their pulses are much broader than the pulses generated by the Ti: sapphire systems [9].

6.1 Parallel systems

Parallel FS microprocessing has three typical systems: multifocus laser writing, processing based on multiple-beam interference, and using a hologram. The first system, the most straightforward technique, is based on a microlens array that splits the single focal spot into multiple foci, uniformly distributed in the focal plane. By the combination of a relay lens and an objective lens, a fabrication resolution comparable with that of the objective lens can be realized. This method can be used for the fabrication of structures with arbitrary 3D geometries in each unit cell. In earlier models, multifocus parallel laser fabrication could only be used for microstructures arranged in periodic arrays. Implementation of multifocus TPP technique, based on individually-controlled phase modulation, enabled rapid prototyping of symmetric and asymmetric complex 2D and 3D structures. The second technique, parallel processing using multiple-beam interference, is potentially faster than the first method and can be used for the fabrication of periodic 2D/3D structures with a wide variety of interference patterns by only a single laser shot. In this technique, an optical diffraction element splits the incident beam into five diverging beamlets, which are collimated by a lens. The phase and amplitude of each beamlet can be tuned, and the beamlets are refocused into the sample by a second lens to create the interference patterns [6].

Various methods have been developed for efficient and flexible fabrication of complex microstructures, including multi-focus parallel processing through optical modulation and diffraction, structured light, photolithography based on a digital micromirror device (DMD), and a liquid crystal spatial light modulator (LC-SLM). The two latter methods, DMD and LC-SLM based FSL processing, can modulate the graphic fabrication of arbitrary structures dynamically with high speed and flexibility, which makes them applicable in various structures. Cross-scale high throughput processing with submicron resolution can also be achieved through a spatiotemporal synchronization focusing approach.

DMD technique is high-throughput, high-contrast, rapid-response, and easy-to-use, appropriate for modulating spatially homogenized flat-headed light sheets into arbitrary 2D patterns, high throughput processing of large-scale microstructural arrays, and single-pulse fabrication of the complex microstructure arrays. It can also be used in combination with digital holography technology for printing various arbitrarily complex 3D structures with high resolution. A larger field of view and patterned area can also be obtained by an objective lens with moderate magnification in applications where super-resolution is not required. A maximum of 44% of laser energy is irradiated on the sample, considering reflection from the DMD surface and energy loss in the gaps between the micromirrors [9].

The second most widely used technique for direct writing is LC-LSM, which modulates the light field phase by a variable distribution of liquid crystals and produces a higher utilization rate of the light energy. This flexible micro-patterning approach was first introduced by researchers from Tokushima University and gained interest in recent years. The accuracy, efficiency, and resolution of LC-SLM-based FSL processing have been improved by achieving the desired patterned beam through holographic algorithm and fabrication approaches. Special microstructures have also been built using structured laser micromachining. It generates Matthew beam, appropriate for fabrication of complex microcages, by regular intensity distribution and the diversification of controllable parameters, as well as Bessel beams, appropriate for fabrication of high aspect ratio microtubes, hollow microhelical structures, and chiral rotating microstructures using the non-diffracting high-quality beam. Both approaches, DMD and LC-SLM based FSL processing, have limitations in achieving high precision and large size simultaneously, for instance, in TPP applications. This limitation has been improved by the spatiotemporal synchronization focusing technology proposed by Cheng et al., which can realize the shortest laser pulse width with the highest laser intensity and improves the efficiency, volume, resolution, and flexibility of FSL fabrication [9].

6.2 Micromachining via interference

FSL has enabled the imprinting of large-scale periodic micro/nanostructures directly on the surface of hard materials or within transparent materials. By adjusting the laser energy, angle of incidence, the number of interference beams, the focal length of the focusing lens, exposure time, laser wavelength, and other parameters, we can modify the period, morphology, and dimension of the structures. Stable and applicable coaxial multi-beam interference can be produced by splitting lenses with a beam splitter and gratings and used for the fabrication of 3D spiral optical fields and chiral microstructures. Michelson interferometer-based FSL spatiotemporal interference approach was proposed for efficient and flexible surface patterning, which can produce a custom-designed gray-scale patterning on a bulk material with a single laser SLM pulse. Direct interference patterning has also been used for nanoparticles size distribution tailoring and multifunctional metal surface modification. This technique, FSL interference, enables the fabrication of periodic functional micro/nanostructures, used in information storage, biomedical engineering, and metamaterials, owing to the invaluable features of this laser technique, including single-step processing, high efficiency, and controllable period [9].

The advantages of FSL have made it appropriate for the fabrication of various optical devices. In the following, we provide a summary of its applications:

- Subtractive fabrication, additive fabrication, and laser-induced modification of microlens arrays. Microlens arrays are used in light modulation, optical sensors, integration of optical systems, parallel micromachining, optical imaging, illumination, and complex quasi-3D surface structures owing to their advantages, including small volume, low cost, and distinctive optical performance. High repetition FSL direct writing enables super-wide-angle imaging, used in completely enclosed cavity concave and convex micro ball lens arrays, embedded in transparent bulk polymer, applicable for aligning with other optical devices.

- Micro/nano gratings, including shape-shifted fiber gratings, highly refractive negative-index gratings, and highly birefringent fiber Bragg gratings. FSL-induced photoreduction of silver ions inside hydrogel can also be used for the fabrication of shrinkable silver diffraction grating.

- Photonic crystals that generate a band gap to prevent the propagation of electromagnetic waves of specific frequencies can be manufactured by FSL, owing to its high precision, high efficiency, special 3D dielectric structure, and non-polluting nature, which enables the fabrication of waveguides, communications, lasers, and exciting molecular chemical reactions. Integration of photonic crystals with all-optical and micro-nanotechnology enables nanotechnology, quantum dot, fiber, and dirac cone. Challenges remain for the fabrication of 3D photonic crystals with bandages.

- Optical fibers: FS direct writing has high efficiency, zero pollution, and 3D precision that makes it appropriate for the processing of small diameter of optic fibers with a transparent material in bulk volume. Multifunctional micro/nanomaterials and devices can also be integrated within a single optical fiber (lab-on-fiber), which has potential for applications in multifunctional sensing and actuating. Both dielectric ad metallic nanostructures can be fabricated on the fiber tip. Direct writing has also been used for the fabrication of multi-core fiber for multiplexing. Multi-components fiber sensors, 3D waveguides, X-couplers, Bragg gratings, microholes, mirrors, optofluidic components, and microfluidic structures can also be processed within a single-mode fiber. Optical fiber using FSL can be used for on-surface and sub-surface fabrication of optical devices and microfluidic devices, which can be integrated onto an optical platform for various materials with diverse physical, chemical, mechanical, and biological properties.

7. Docking

For the fs lasers to be docked with the eyes, head must be in the fixed laser housing, the articulated arm holding handpiece be on the eye, and there should be a distance between eye surface and laser optics. Sterile, single-use components, referred to as "patient interfaces," are used to actually make contact with the eye. There are two types of interfaces in use: an applanating interface, which has a curved or flat surface that touches the cornea directly; and a liquid-filled interface, which has a vacuum ring that makes contact with the sclera or the outer cornea and a liquid-filled center. The liquid-filled interface maintains the cornea's natural shape while enabling laser energy transmission. Despite the fact that contact surfaces momentarily alter the curvature of the cornea [10], mechanical contact significantly stabilizes the cornea during surgery. This is especially crucial in refractive surgery, where micrometer-level tissue movement must be prevented and precise incisions are necessary. Because refractive surgery has no obvious clinical downsides [11–13], contact interfaces will predominate in corneal surgery in the future (**Figure 6**). Liquid-filled interfaces that cause minimum disruption to the eye may end up being the best option for cataract surgery. Vacuum docking contact stability during laser emission is crucial. Loss of touch can lead to wrong-plane cuts. All lasers are designed to automatically monitor vacuum levels and halt laser emission upon loss of contact. Eye surgeons watch their patients during the process and might manually interrupt or pause it if difficulties arise. In laser systems with an articulated arm, surgeons can use their manual abilities to actively stabilize the laser handpiece. With a vacuum loss, treatment can be restarted after a new docking.

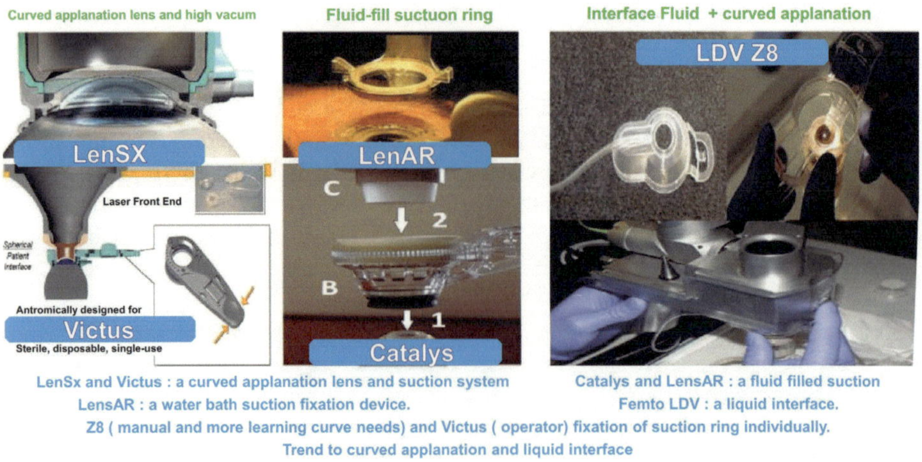

Figure 6.
Docking by femtosecond laser.

8. Applications in ophthalmology

As described above, FSL has several advantages, such as minimum collateral damage, which enables bladeless high-precision incisions within the tissue by different patterns and depths. Accordingly, FSL has been implemented in several ophthalmologic surgical procedures, which has substantially revolutionized the safety, precision, and predictability of surgeries (**Figure 7**). The ongoing advances in FSL technology since its introduction are making the newer machines with improved postsurgical outcomes in terms of visual quality, stability of treatments, and complications [14].

Fs-laser cuts can be used to produce “pockets” in the cornea, from which either material can be extracted or implanted, in a variety of refractive surgeries (**Figure 8**). The cornea’s refracting power is altered in both scenarios.

8.1 Refractive surgery

8.1.1 Ablative procedure

FSLs are used in various refractive surgery procedures such as FSL-assisted laser in LASIK, refractive lenticular extraction (RELEX), and small-incision lenticule extraction (SMILE) (**Figure 9**). During LASIK procedure, first, a corneal flap is created by FSL; for this purpose, the laser spot moves inside the cornea at a programmed thickness, and by a perfect cut, the lamella is created. Then, the FS bundle continues along the flap circumference, and the flap is lifted; then, excimer- or solid-state UV-laser energy is used to change the cornea’s refractive power by flattening or steepening the stromal bed. Later, the flap is repositioned.

Advances in FS laser technology, including the availability of fast-firing rates, enabled smaller spot sizes and tighter spot separation, which made sweeping and lifting of the flap easier. A well-fitting manhole is also enabled by the vertical side cuts of FSL. Compared with the traditional photo-refractive keratectomy, LASIK

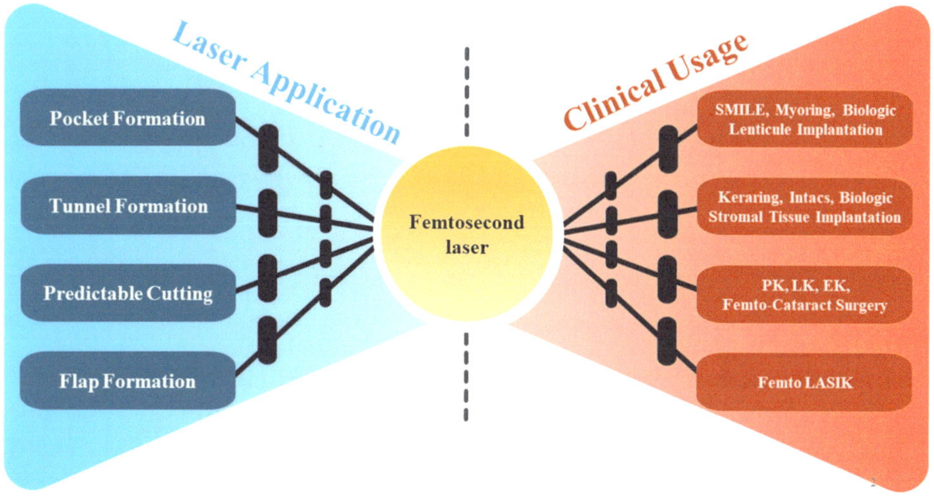

Figure 7.
Considerable applications of different techniques of femtosecond laser (Left) and outstanding clinical usages (Right); SMILE: Small Incision Lenticule Extraction; PK: Penetrating Keratoplasty; LK: Lamellar Keratoplasty; EK: Endothelial Keratoplasty; Femto LASIK: Femtosecond-assisted (Femto) Laser In-Situ Keratomileusis (LASIK).

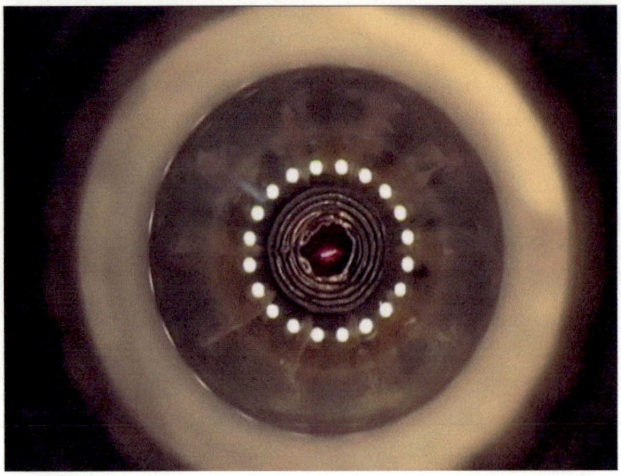

Figure 8.
Femtosecond laser cut.

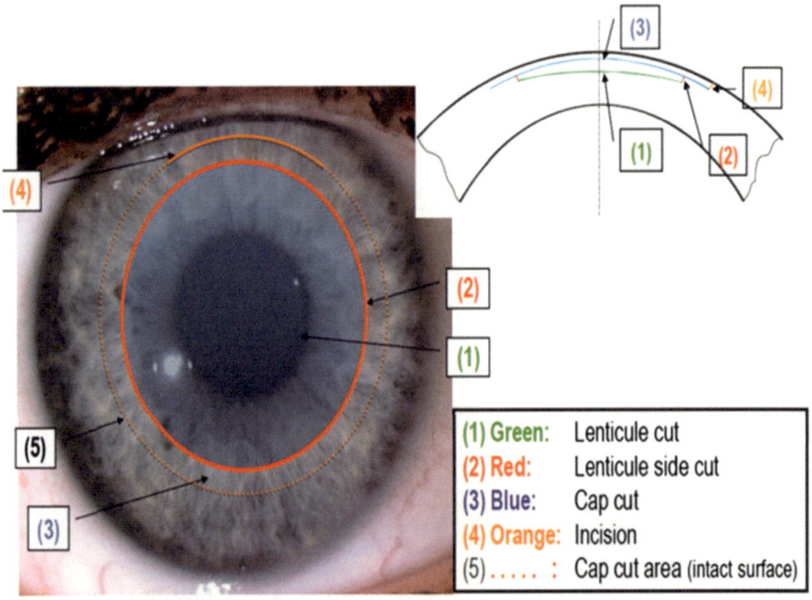

Figure 9.
Lenticule cut via femtosecond laser.

(using IntraLase™) showed to have a shorter rehabilitation period, more predictable flap thickness, better astigmatic neutrality, less epithelial injury, and faster stabilization of visual outcome; but is associated with a higher risk of dry eye and flap-related complications, such as transient light-sensitivity syndrome after surgery and diffuse lamellar keratitis in the flap interface. Nonetheless, the risk of dry eye and visual loss (related to intraocular pressure) in FS-LASIK is less than the conventional microkeratome-assisted LASIK. In lenticule extraction, FSL system is used for cutting a small lenticule in corneal tissue, and similar to the flap created in LASIK, the flap is lifted, the lenticule is removed, and then the flap is repositioned. Also, in the SMILE

8.3 Cataract surgery

In cataract surgery, the current FS systems allow imaging and measurement of the anterior segment of the eye, cutting the tissue at the desired location depth, pattern, and size, full-thickness corneal incisions for the introduction of instruments to the eye, partial thickness corneal incisions for treatment of corneal astigmatism, circular incision to the anterior lens capsule (capsulotomy), and fragmentation of the cataractous lens nucleus (**Figure 11**). Advantages of FSL-assisted cataract surgery include precision and repeatability of incisions, lower ultrasound energy used for lens nucleus emulsification/liquefication), perfect sizing, and predictability of corneal incisions and capsulotomy (**Figure 12**). Despite the numerous studies suggesting the superiority of different FSLs used for cataract surgery (**Table 1**) over the conventional phacoemulsification manual operation and emphasis of review studies on the advantages of FSL method, meta-analysis studies failed to prove its superiority considering the overall outcome of patients. The introduction of low-energy lasers may have improved outcomes compared with first-generation FSL [5, 14].

Despite the above-mentioned clear benefits and multiple studies indicating superiority of femto laser cataract surgery in completing the single surgical stages [22, 23]. Review papers stress fs-assisted cataract surgery in specific patient groups, i.e., those with low corneal endothelial cell numbers, but a clear advantage of the fs approach over manual phacoemulsification is not found in normal cases [24, 25]. Primary posterior capsulotomy assisted by fs laser, as well lens capsule labeling, altering the power of the intraocular lens (IOL) postoperatively using more mobile, and adaptable fs-laser systems will make the future of femtosecond laser brighter (**Figure 13**) [26].

Figure 11.
History and Evolution of Femtosecond Laser Assisted Cataract Surgery (FLACS.

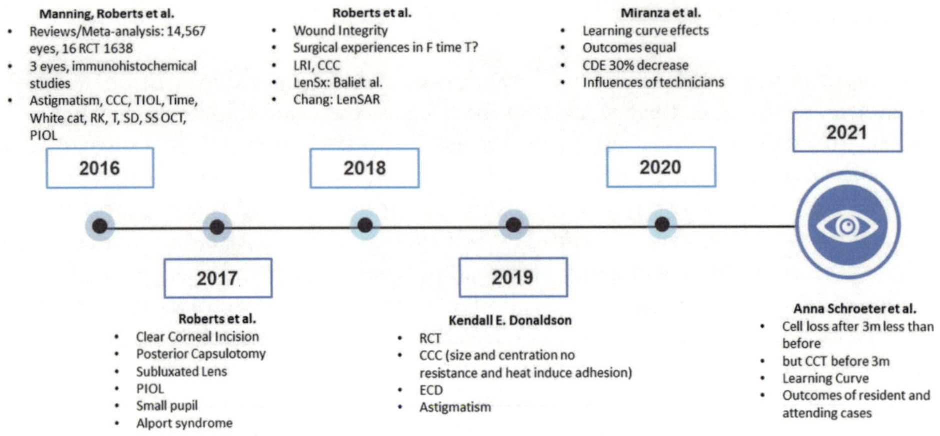

Figure 12.
Femtosecond Laser Assisted Cataract Surgery (FLACS) vs. Micro Incision Cataract Surgery (MICS); Effects of Learning Curve.

Femto-laser	CATALYS	LenSx	LENSAR	VICTUS
Arcuate incisions	Surface and Intrastromal	Surface and Intrastromal	Surface and Intrastromal	Capable of surface or stromal
Pulse frequency (kHz)	120	50	80	Up to 160
FDA approvals	Corneal + arcuate incisions Ant capsulotomy Lens fragmentation	Corneal + arcuate incisions Ant capsulotomy Lens fragmentation Corneal flap	Corneal + arcuate incisions Ant capsulotomy Lens fragmentation	Corneal + arcuate incisions Ant capsulotomy Corneal flap
Patient interface design	Liquid optics Non-applanating Liquid interface 2-piece Vacuum docking	Soft fit Applanating Curved lens 1-piece Vacuum docking	Robocone Non-applanating Fluid interface 2-pieces Vacuum docking	Dual modality Curve lens Applanating Spherical 2-pieces Vacuum docking Solid + Liquid
Docking	Ocular surface bathed in saline solution No corneal applanation No glaucoma contraindication	Curved applanation No glaucoma Intra indication (soft fit PI)	No corneal applanation	Soft docking for capsulotomy and lens fragmentation Regular docking for corneal incisions
IOP rise	10.3 mm Hg rise	16.4 mm Hg rise (cionni, ASCRS 2012 presentation)	Unknown (currently under evaluation) Automatic (augmented reality camera)	Unknown (currently under evaluation) Manual

Femto-laser	CATALYS	LenSx	LENSAR	VICTUS
Imaging type	3D spectral domain OCT Video microscope and FS Laser to enable image-guided cataract surgery	3D spectral domain OCT Video microscope and FS Laser to enable image-guided cataract surgery	3D ray-tracing Confocal Structural Illumination	3D spectral domain OCT Video microscope and FS Laser to enable image-guided cataract surgery

Table 1.
The commercial machines for FLACS.

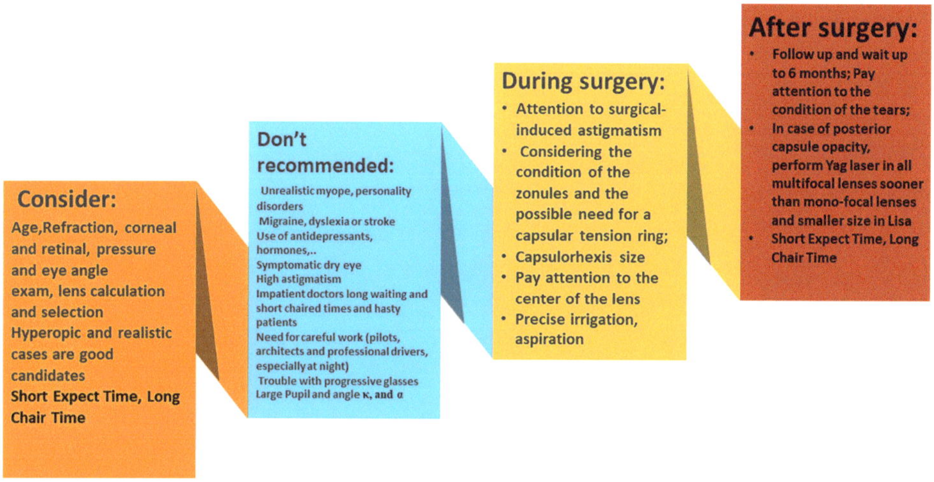

Figure 13.
Femtosecond Laser-Assisted Cataract Surgery (FLACS).

9. Conclusions

As a precise and reliable tool in the field of ophthalmic surgery, fs-laser technology has evolved over the past few decades. Fs-laser-assisted cataract and corneal surgery have reached a high standard worldwide, despite the fact that some applications are still in the early stages of development. These procedures have been made safer and more effective by the use of fslaser technology, which has opened the door to new surgical approaches.

Conflict of interest

The authors declare no conflict of interest.

Author details

Sana Niazi[1] and Farideh Doroodgar[2,3]*

1 Medical Research center, Shahid Beheshti University of Medical Sciences, Tehran, Iran

2 Translational Ophthalmic Research Cente, Tehran University of Medical Science, Tehran, Iran

3 Negah Specialty Ophthalmic Research Center, Shahid Beheshti University of Medical Sciences, Tehran, Iran

*Address all correspondence to: farinaz_144@yahoo.com

References

[1] Sibbett W, Lagatsky A, Brown C. The development and application of femtosecond laser systems. Optics Express. 2012;**20**(7):6989-7001

[2] Vogel A et al. Mechanisms of femtosecond laser nanosurgery of cells and tissues. Applied Physics B. 2005;**81**(8):1015-1047

[3] Guo B et al. Femtosecond laser micro/nano-manufacturing: Theories, measurements, methods, and applications. Nanomanufacturing and Metrology. 2020;**3**(1):26-67

[4] Soong HK, Malta JB. Femtosecond lasers in ophthalmology. American Journal of Ophthalmology. 2009;**147**(2):189-197 e2

[5] Asshauer T et al. Femtosecond lasers for eye surgery applications: Historical overview and modern low pulse energy concepts. Advanced Optical Technologies. 2021;**10**(6):393-408

[6] Sugioka K, Cheng Y. A tutorial on optics for ultrafast laser materials processing: Basic microprocessing system to beam shaping and advanced focusing methods. Advanced Optical Technologies. 2012;**1**(5):353-364

[7] Jiang, L, Tsai H-L. Improved two-temperature model and its application in ultrashort laser heating of metal films. Journal of Heat Transfer. 2005;**127**(10):1167-1173

[8] Mao S et al. Dynamics of femtosecond laser interactions with dielectrics. Applied Physics A. 2004;**79**(7):1695-1709

[9] Wang X et al. Femtosecond laser-based processing methods and their applications in optical device manufacturing: A review. Optics & Laser Technology. 2021;**135**:106687

[10] Mirshahi A, Kohnen T. Effect of microkeratome suction during LASIK on ocular structures. Ophthalmology. 2005;**112**(4):645-649

[11] Flaxel CJ et al. Proposed mechanism for retinal tears after LASIK: An experimental model. Ophthalmology. 2004;**111**(1):24-27

[12] Kanclerz P, Grzybowski A. Does corneal refractive surgery increase the risk of retinal detachment? A literature review and statistical analysis. Journal of Refractive Surgery. 2019;**35**(8):517-524

[13] Toth CA et al. LASIK and vitreous pathology after LASIK. Ophthalmology. 2002;**109**(4):624

[14] Aristeidou A et al. The evolution of corneal and refractive surgery with the femtosecond laser. Eye and Vision. 2015;**2**(1):1-14

[15] Roszkowska AM, et al. Use of the femtosecond lasers in ophthalmology. In: EPJ Web of Conferences; EDP Sciences. 2018

[16] Riau AK et al. Stromal keratophakia: Corneal inlay implantation. Progress in Retinal and Eye Research. 2020;**75**:100780

[17] Seitz B et al. Suppl-1, M7: Penetrating keratoplasty for keratoconus–excimer versus femtosecond laser trephination. The Open Ophthalmology Journal. 2017;**11**:225

[18] Seitz B et al. Reconsidering sequential double running suture

removal after penetrating keratoplasty: A prospective randomized study comparing excimer laser and motor trephination. Cornea. 2018;**37**(3):301-306

[19] Reinhart WJ et al. Deep anterior lamellar keratoplasty as an alternative to penetrating keratoplasty a report by the american academy of ophthalmology. Ophthalmology. 2011;**118**(1):209-218

[20] Anwar M, Teichmann KD. Big-bubble technique to bare Descemet's membrane in anterior lamellar keratoplasty. Journal of Cataract & Refractive Surgery. 2002;**28**(3):398-403

[21] de Macedo JP et al. Femtosecond laser-assisted deep anterior lamellar keratoplasty in phototherapeutic keratectomy versus the big-bubble technique in keratoconus. International Journal of Ophthalmology. 2018;**11**(5):807

[22] Wang J et al. Intra and post-operative complications observed with femtosecond laser-assisted cataract surgery versus conventional phacoemulsification surgery: A systematic review and meta-analysis. BMC Ophthalmology. 2019;**19**(1):177

[23] Day AC et al. Laser-assisted cataract surgery versus standard ultrasound phacoemulsification cataract surgery. Cochrane Database of Systematic Reviews. 2016;7:CD010735

[24] Kanclerz P, Alio JL. The benefits and drawbacks of femtosecond laser-assisted cataract surgery. European Journal of Ophthalmology. 2021;**31**(3):1021-1030

[25] Hooshmand J, Vote BJ. Femtosecond laser-assisted cataract surgery: technology, outcome, future directions, and modern applications. The Asia-Pacific Journal of Ophthalmology. 2017;**6**(4):393-400

[26] Dick HB. Future perspectives of the femtosecond laser in anterior segment surgery. Der Ophthalmologe: Zeitschrift der Deutschen Ophthalmologischen Gesellschaft. 2020;**117**(5):431-436

Chapter 3

Modeling of Laser-Irradiated Biological Tissue

Sumit Kumar

Abstract

The laser has been widely used in medical fields. One application of the laser is laser-based photo-thermal therapy, wherein the short-pulsed laser is generally used to destroy the cancerous cells. The efficacy of the laser-based photo-thermal therapy can be improved if we minimize the thermal damage to the surrounding healthy tissue. So, it is essential to understand the laser-tissue interaction and thermal behavior of biological tissue during laser-based photo-thermal therapy. The light propagation through the biological tissue is generally mathematically modeled by the radiative heat transfer equation (RTE). The RTE has been solved using the discrete ordinate method (DOM) to determine the intensity inside the laser-irradiated biological tissue. Consequently, the absorbed photon energy act as the source term in the Fourier/non-Fourier model-based bio-heat transfer equation to determine the temperature distribution inside the biological tissue subjected to short-pulse laser irradiation. The non-Fourier model-based bio-heat transfer equation is numerically solved using the finite volume method (FVM). The numerical results have been compared with the analytical results obtained using the finite integral transform (FIT) technique. A comparative study between the Fourier and non-Fourier heat conduction models has also been carried out.

Keywords: laser-based photo-thermal therapy, bio-heat transfer, transient radiative transfer equation, non-Fourier heat conduction model, numerical study

1. Introduction

Cancer is a group of diseases in which uncontrolled growth and invade of abnormal cells into other parts of the body [1]. The World Health Organization says nearly one in six died from cancer in 2020 [2]. Thus, it is essential to detect the cancerous cells at their early stage and destroy them to minimize possible damage to the surrounding healthy tissue. So, various diagnosis techniques such as computerized tomography (C.T.) scan, magnetic resonance imaging (MRI), ultrasound, X-ray, positron emission tomography (PET) scan, etc., and treatment such as surgery, chemotherapy, laser-based photo-thermal therapy, etc., have been developed in the past.

Laser-based photo-thermal therapy has gained unprecedented growth among all available treatment techniques in the past few decades. In this therapy, the temperature of the tissue is increased above the pre-defined threshold value using the laser, which destroys the cancerous cells [3]. In other words, the photon energy has been absorbed by the laser-irradiated biological tissue, which subsequently increases its internal energy; as a result, the temperature of the tissue increases. Amongst the various types of laser, the short-pulse laser has gained substantial importance because it can transmit high energy in a very short interval of time (order of femtosecond to picosecond) [3]. This advantage helps in increasing the temperature to the desired level in the confined region. Therefore, the short-pulse laser has been usually used to destroy cancerous cells during laser-based photo-thermal therapy. It is worth noting that tissue gets thermally damaged when its temperature crosses 43°C [4]. So, the major challenge in the medical field is to destroy the cancerous cells without damaging the surrounding healthy tissue. Thus, it is essential to understand the thermal characteristics of the laser-irradiated biological tissue to improve the efficacy of laser-based photo-thermal therapy.

The modeling of laser-tissue interaction is challenging because it is absorbing as well as scattering in nature due to its constituents such as water, hemoglobin, melanin RBCs, cell membrane, etc. [3, 5]. So, Various mathematical models such as Beer-Lambert's law, diffusion approximation theory, radiative transfer equation (RTE), etc., have been developed to model the laser-tissue interaction. Among these mathematical models, the Beer-Lambert's law ($I = I_o \exp(-\kappa z)$)is the simple and relatively straightforward mathematical model, which is solved by the researchers to determine the intensity (I) variation inside the laser-irradiated biological tissue [6, 7]. Here, κ represents the absorption coefficient. However, the limitation of this model is that it assumes that the biological tissue is purely absorbing in nature. Few researchers used the diffusion approximation theory to determine the intensity distribution by considering that the biological tissue is highly scattering in nature [8, 9]. So, assuming the biological tissue is either purely absorbing or highly scattering misleads the results. Therefore, the researchers [4] have suggested using the RTE (Eq. 1), which is the most appropriate mathematical model for modeling the laser-tissue interaction [10].

$$\frac{1}{c}\frac{\partial I}{\partial t} + \frac{dI}{ds} = -\beta I + \kappa I_b + \frac{\sigma_s}{4\pi}\int_{4\pi} I\varnothing(\Omega, \Omega')d\Omega' \qquad (1)$$

where, c,β,σ_s and $\varnothing$ represents the speed of light in the medium, extinction coefficient (sum of scattering and absorption coefficient), scattering coefficient, and scattering phase function, respectively. The first and second term on the left-hand side of Eq. (1) represents the temporal and spatial variation of intensity, respectively. On the other hand, the first, second, and third terms on the right-hand side of Eq. (1) denote the attenuation of intensity due to the absorption and out-scattering, augmentation of intensity due to the emission and in-scattering, respectively.

Eq. (1) is the integrodifferential equation, and getting the analytical solution is quite challenging due to the presence of the in-scattering terms (third term of the right-hand side of Eq. (1)). So, researchers have developed various numerical methods, such as the discrete transfer method (DTM) [11], finite volume method (FVM) [12], discrete ordinate method (DOM) [13], etc., to convert this integrodifferential equation into a simpler form to get the numerical solution. The various researchers performed the comparative studies between these numerical methods and found that the DOM is computationally most efficient, uses comparatively lesser memory and the effective wave speed predicted by DOM comes closer to the actual speed of light [14–16]. So, the DOM has been employed in the present study to solve the RTE.

To understand the thermal behavior of the laser-irradiated biological tissue, we need to solve the bio-heat transfer equation (Eq. 2) wherein the energy generated by the laser acts as the source terms.

$$\rho c_v \frac{\partial T}{\partial t} = -\nabla \bullet \vec{q} + \omega_b \rho_b c_b (T_b - T) + Q_m - \nabla \bullet \vec{q_r} \tag{2}$$

where ρ, c_v, ω_b, ρ_b, and c_b denote the density of the tissue, specific heat of the tissue, blood perfusion, density of blood, and specific heat of blood, respectively.

The term on the left-hand side of Eq. (2) represents the time rate of change of energy inside the biological tissue. On the other hand, the first, second, third, and fourth terms on the right-hand side of Eq. (2) represent the diffusion term due to the net heat transfer across the boundaries, blood perfusion term due to the energy exchange between the tissue and surrounding capillary blood vessels, metabolic heat generation term, and the divergence of the radiative heat flux which represents net radiative heat loss from the medium.

When Fourier's law of heat conduction (Eq. 3) is substituted into Eq. (2), the resulting equation is known as the Fourier model-based bio-heat transfer equation, also known as Pennes bio-heat transfer equation.

$$\vec{q} = -k\nabla T \tag{3}$$

The limitation of the Pennes bio-heat transfer equation is that it assumes that arterial blood temperature is almost uniform throughout the tissue domain while the venous blood temperature equals that of the local tissue temperature [17]. However, the assumption of thermal equilibrium between the blood vessel and the surrounding tissue is not valid when the diameter of the blood vessel is more than 500 μm (i.e., large blood vessel). In that case, the energy equation has been separately solved for the tissue and blood vessel domain to get the accurate temperature distribution. The interested reader can find the numerical solution of the energy equation to determine the temperature distribution inside the laser-irradiated biological tissue embedded with the large blood vessel elsewhere [18].

Few researchers experimentally demonstrated that Fourier's law of heat conduction under predicts the temperature value that does not accurately match the experimental data because of the inherent assumption of the infinite speed of thermal wave propagation through the biological tissue [4, 19, 20]. Furthermore, the Fourier model gives inaccurate results in an area where used the short-pulse laser, the temperature is in the range of cryogenic, studying the thermal response of non-homogeneous structures, e.g., biological tissue [21]. To overcome the limitations of Fourier heat conduction models, researchers have modified Eq. (3) by considering the relaxation time associated with the heat flux (τ_q) and temperature gradient (τ_T), and it is known as the dual-phase lag (DPL) model (Eq. 4).

$$\vec{q} + \tau_q \frac{\partial \vec{q}}{\partial t} = -k\left\{\nabla T + \tau_T \frac{\partial(\nabla T)}{\partial t}\right\} \tag{4}$$

When τ_T is equal to zero, the DPL model becomes the Cattaneo-Vernotte (C-V) equation. However, the limitation of this model is that it doesn't consider the effects of microstructural interactions in a non-homogeneous medium, e.g., biological samples [21]. The DPL model becomes the conventional Fourier's law of heat conduction when both relaxation times are equal to zero. So, the DPL model is the generalized form of the non-Fourier heat conduction model. Thus, the numerical and analytical solutions of the DPL model-based bio-heat transfer equation have been obtained using the FVM and finite integral technique (FIT), respectively, which are discussed in Sections 3 and 4 respectively.

As mentioned earlier, the last term on the right-hand side of Eq. (2) represents the divergence of radiative heat flux (Eq. (5)), which is obtained by solving the Eq. (1).

$$\nabla \bullet \overrightarrow{q_r} = \kappa(4\pi I_b - G) = \kappa\left(4\sigma T^4 - G\right) \tag{5}$$

where, I_b and $G = \sum_{m=1}^{M} \omega^m I^m$ represent the black body intensity and the incident intensity. Here ω^m denotes the angular weight in the particular direction m, and M is the total number of divisions of the 4π solid angle.

It is to be noted here that the time scale used for solving the RTE is the order of picosecond because the light propagates through the biological tissue generally has the same order of time scale. However, at such time scales, the contribution of the diffusion term, blood perfusion term, and metabolic heat generation term (the first three terms on the right-hand side of Eq. (2)) to the increment of the temperature of the tissue are insignificant as compared to the divergence of the radiative heat flux. However, the time scale of the order of the millisecond is generally used for solving the bio-heat transfer equation to capture the effect of these terms. Therefore, it is a multi-time scale problem because the two different time scales have been used to solve the RTE and bio-heat transfer equation. The solution procedure for solving this multi-time scale problem has been discussed in Section 2.

2. Problem formulation

A two-dimensional squared-shaped biological tissue having the dimensions of 2 mm × 2 mm was considered in the present study, which is shown in **Figure 1**. The short-pulse laser is vertically incident at the top surface of the domain. Short laser pulses have been modeled as Gaussian profile in the space domain and Top-Hat profile in the time domain with a pulse width of t_p. When light propagates through the biological tissue, the intensity has two components: collimated and diffuse. The collimated component of the intensity has been mathematically expressed by Eq. (6) [21].

$$\begin{aligned} I_c(x, y, \Omega, t) = I_{cmax} \times exp\left\{-(x-L/2)^2/d^2\right\} \times exp\{-\beta(W-y)/|\eta_c|\} \\ \times\left[H\{\beta ct - \beta(W-y)/|\eta_c|\} - H\{(\beta ct - \beta(W-y)/|\eta_c|) - \beta ct_p\}\right] \\ \times \delta(\mu-\mu_c)\delta(\eta-\eta_c) \end{aligned} \tag{6}$$

where, I_{cmax}, L, W, μ, and η represent the maximum intensity incident on the top surface of the tissue phantom, the length of tissue, the width of the tissue, and direction cosines in x- and y-directions, respectively. Here, δ denotes the Dirac-delta function, which considers the angle of incidence, and H is the Heaviside function which ensures that the laser energy is available at any location only for the pulse width of the laser. The first and second exponential terms in Eq. (6) represent the Gaussian profile in the space domain and the attenuation of collimated intensity due to absorption and out-scattering, respectively.

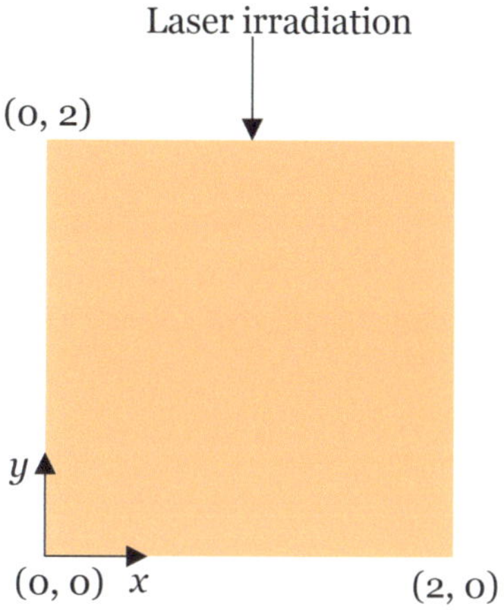

Figure 1.
Schematic diagram of the physical domain under consideration (dimensions in mm).

The diffuse component of the intensity is obtained by solving the RTE. So, Eq. (1) is simplified for the two-dimensional rectangular domain and also neglects the emission from the tissue phantom because the intensity of the short-pulse laser used for irradiating the sample is much higher than the blackbody radiation intensity:

$$\frac{1}{c}\frac{\partial I_d}{\partial t} + \mu\frac{\partial I_d}{\partial x} + \eta\frac{\partial I_d}{\partial y} = -\beta I_d + \frac{\sigma_s}{4\pi}\int_{4\pi} I\varnothing(\Omega, \Omega')d\Omega' \tag{7}$$

The boundary conditions for a diffusely emitting and reflecting wall at a given point r_w on that surface can be expressed as [21]:

$$I(r_w, \hat{s}, t) = \in(r_w)I_b(r_w, t) + \frac{\rho(r_w)}{\pi}\int_{\hat{n}\bullet\hat{s}'<0} I(r_w, \hat{s}', t)|\hat{n}\bullet\hat{s}'|d\Omega', \hat{n}\bullet\hat{s} > 0 \tag{8}$$

Here $\hat{n}$ is the inward normal vector to the surface. The first and second terms on the right-hand side of Eq. (8) represent the emission and reflection from the wall, respectively. The expression for boundary conditions given by Eq. (8), in general, holds valid for all the four walls of the rectangular enclosure.

Once the intensity distribution inside the laser-irradiated biological tissue is obtained, then the divergence of the radiative heat flux (Eq. 5) is calculated. As earlier mentioned, finding the temperature distribution inside the biological tissue subjected to the short-pulse laser is a multi-time scale problem. So, we need to divide the Eq. (2) into Eqs. (9) and (10). Eq. (9) is solved for determining local temperature rise due to a single pulse for which the time scale is the order of picoseconds.

$$\rho c_v\frac{\partial T}{\partial t} = -\nabla\bullet\overrightarrow{q_r} \tag{9}$$

To determine the temperature distribution inside the biological tissue subjected to a train of pulses, the temperature rise due to the single pulse is added to the temperature distribution at the previous time instant if the time difference between the two consecutive pulses is equal to the time step employed else it is not added. The temperature distribution at the previous time instant is used as the initial condition to determine the temperature distribution at the current time instant by solving the Eq. (10) [21]. It is to be noted that the time scale used for solving Eq. (10) is the order of milliseconds.

$$\rho c_v \frac{\partial T}{\partial t} = -\nabla \bullet \vec{q} + \omega_b \rho_b c_b (T_b - T) + Q_m \tag{10}$$

The DPL model-based bio-heat transfer equation has been solved in the present study, which is obtained by eliminating temperature from Eq. (10) using the Eq. (4), and the governing equation is written in terms of heat flux which is expressed as:

$$\frac{\tau_q}{\alpha} \frac{\partial^2 \vec{q}}{\partial t^2} + \frac{1}{\alpha} \frac{\partial \vec{q}}{\partial t} = \nabla \left(\nabla \bullet \vec{q} \right) + \tau_T \frac{\partial \left[\nabla \left(\nabla \bullet \vec{q} \right) \right]}{\partial t} \tag{11}$$

It is to be noted here that the blood perfusion and metabolic heat generation terms have been neglected while deriving Eq. (11) because their contribution is negligible to temperature rise [21].

The required boundary conditions for solving the Eq. (10) are as follows: the all three walls except the top wall (shown in **Figure 1**) are maintained at the core body temperature (37°C), while the top wall is subjected to the convective boundary conditions (heat transfer coefficient of 15 $W/m^2 \bullet K$ and ambient temperature of 25°C) [21]. The initial condition (t = 0) for temperature is equal to 37°C, and the two initial conditions are required for Eq. (11) because it has a second-order derivative concerning the time, so heat flux and time derivative of heat flux are assumed to be zero.

Eqs. (10) and (11) have been numerically solved using the FVM, and its solution is given in Section 3.

The following laser parameters have been used in this study: beam diameter (d)= 0.025 mm, amplitude of pulse (I_{cmax})= 1.6×10^{-3} $J/mm^2/ps$, pulse width (t_p)= 4.6667 ps, repetition rate (fr)= 1 kHz and wavelength= 1100 nm [21]. The absorption coefficient (κ) and scattering coefficient (σ_s) of tissue are 0.051 mm^{-1} and 6.14 mm^{-1}, respectively [4, 21]. The thermo-physical properties of the tissue are as follows: density (ρ)= 1000 kg/m^3, thermal conductivity (k)= 0.63 W/m•K and specific heat (c_v)= 4200 J/kg•K [21].

3. Numerical modeling

This section discusses the methodology for solving the RTE (Eq. 7) using the DOM. Then, the numerical solution of the DPL model-based bio-heat transfer is obtained using the FVM to determine the temperature distribution inside the laser-irradiated biological tissue.

In DOM, the RTE, an integrodifferential equation, has been transformed into a set of partial differential equations [10]. So, the RTE for the diffuse component of intensity (Eq. (12)) with the corresponding boundary conditions (Eq. 13) in any given discrete direction ($\Omega^{m,n}$) can be written as follows

$$\frac{1}{c} \frac{\partial I_d^{m,n}}{\partial t} + \mu^{m,n} \frac{\partial I_d^{m,n}}{\partial x} + \eta^{m,n} \frac{\partial I_d^{m,n}}{\partial y} = -\beta I_d^{m,n} + \frac{\sigma_s}{4\pi} \sum_{m'=1}^{N_\theta} \sum_{n'=1}^{N_\varphi} I_d^{m',n'} \varnothing^{m',n';m,n} \omega_\theta^{m'} \omega_\varphi^{n'} + \frac{\sigma_s}{4\pi} I_c^{m_c,n_c} \varnothing^{m_c,n_c;m,n} \tag{12}$$

$$\text{Left wall } (x = 0,\ 0 < y < W) : I_d^{m,n} = 0, \mu^{m,n} > 0 \tag{13}$$

$$\text{Right wall } (x = L,\ 0 < y < W) : I_d^{m,n} = 0, \mu^{m,n} < 0 \tag{14}$$

$$\text{Bottom wall } (y = 0,\ 0 < x < L) : I_d^{m,n} = 0, \eta^{m,n} > 0 \tag{15}$$

$$\text{Top wall } (y = W,\ 0 < x < L) : I_d^{m,n} = 0, \eta^{m,n} < 0 \tag{16}$$

The boundary conditions Eqs. (13)–(16) have been obtained using Eq. (8) under the assumption that the boundary is non-reflecting and the magnitude of the light intensity emitted from the surface is negligible compared to the intensity of the short-pulse laser used.

Using a fully implicit backward differencing scheme in time, Eq. (12), after simplification, becomes

$$\mu^{m,n}\frac{\partial I_d^{m,n}(t^*)}{\partial x} + \eta^{m,n}\frac{\partial I_d^{m,n}(t^*)}{\partial y} + \left[\frac{\beta}{\Delta t^*} + \beta\right] I_d^{m,n}(t^*) = \frac{\beta}{\Delta t^*} I_d^{m,n}(t^* - \Delta t^*) + S_t^{m,n} \tag{17}$$

where $t^* = \beta c t$,

$$S_t^{m,n} = \frac{\sigma_s}{4\pi}\sum_{m'=1}^{N_\theta}\sum_{n'=1}^{N_\varphi} I_d^{m',n'} \varnothing^{m',n';m,n} \omega_\theta^{m'} \omega_\varphi^{n'} + \frac{\sigma_s}{4\pi} I_c^{m_c,n_c} \varnothing^{m_c,n_c;m,n} \tag{18}$$

Integrating Eq. (17) over the control volume ΔV leads to

$$\begin{aligned} &\mu^{m,n}\Delta y\left[I_{d,E}^{m,n}(t^*) - I_{d,W}^{m,n}(t^*)\right] + \eta^{m,n}\Delta x\left[I_{d,N}^{m,n}(t^*) - I_{d,S}^{m,n}(t^*)\right] \\ &= -\left[\frac{\beta}{\Delta t^*} + \beta\right]\Delta V I_{d,P}^{m,n}(t^*) + \Delta V S_{t,P}^{m,n} + \frac{\beta\Delta V}{\Delta t^*} I_{d,P}^{m,n}(t^* - \Delta t^*) \end{aligned} \tag{19}$$

where, $I_{d,P}^{m}$ is the intensity at the cell center P, $S_{t,P}^{m}$ is the source term at the cell center P and $\Delta V = \Delta x \times \Delta y$. For relating the cell-surface intensities with cell-center intensity, the step differencing scheme has been used to avoid any possibility of providing unphysical results [10].

Eq. (19) is simplified using the step differencing scheme, and the resulting equation becomes as:

$$I_{d,P}^{m,n}(t^*) = \frac{\left[\mu^{m,n}\Delta y I_{d,W}^{m,n}(t^*) + \eta^{m,n}\Delta x I_{d,S}^{m,n}(t^*) + \Delta V S_{t,P}^{m,n} + \left(\frac{\beta\Delta V}{\Delta t^*}\right) I_{d,P}^{m,n}(t^* - \Delta t^*)\right]}{\mu^{m,n}\Delta y + \eta^{m,n}\Delta x + \left[\frac{\beta}{\Delta t^*} + \beta\right]\Delta V} \tag{20}$$

As earlier mentioned, the solution of RTE provides the intensity distribution, which is used to determine the divergence of the radiative heat flux (Eq. 5). Subsequently, find the temperature rise due to a single pulse by solving the Eq. (9). Then, Eq. (10) has been solved using the FVM to determine the temperature distribution inside the biological tissue subjected to the train of pulses. The algorithm for solving this multi-time scale problem has been discussed in Section 2.

Under the Cartesian coordinates system, the governing equation (Eq. 11) in x- directions can be expressed as [21]

$$\frac{\tau_q}{\alpha}\frac{\partial^2 q_x}{\partial t^2} + \frac{1}{\alpha}\frac{\partial q_x}{\partial t} = \frac{\partial}{\partial x}\left(\frac{\partial q_x}{\partial x} + \frac{\partial q_y}{\partial y}\right) + \tau_T\frac{\partial}{\partial t}\left\{\frac{\partial}{\partial x}\left(\frac{\partial q_x}{\partial x} + \frac{\partial q_y}{\partial y}\right)\right\} \tag{21}$$

Integrating Eq. (21) over a given control volume and over the time step of Δt, i.e., from t to $t+\Delta t$ followed by discretization based on the backward difference in time leads to the following discretized form of the Eq. (22) [21].

$$a_{xP} q_{xP}^{t+\Delta t} = a_W q_{xW}^{t+\Delta t} + a_E q_{xE}^{t+\Delta t} + b_x \tag{22}$$

where $a_W = \frac{\Delta y}{(\delta x)_w}(\Delta t + \tau_T)$, $a_E = \frac{\Delta y}{(\delta x)_e}(\Delta t + \tau_T)$, $a_{xP} = \frac{\Delta x \Delta y}{\alpha}\left(1 + \frac{\tau_q}{\Delta t}\right) + a_W + a_E$,

$$\begin{aligned} b_x = {} & \left(\frac{\Delta x \Delta y}{\alpha}\left(1 + \frac{2\tau_q}{\Delta t}\right) + \frac{\Delta y \tau_T}{(\delta x)_w} + \frac{\Delta y \tau_T}{(\delta x)_e}\right) q_{xP}^{t} - \frac{\Delta y \tau_T}{(\delta x)_w} q_{xW}^{t} - \frac{\Delta y \tau_T}{(\delta x)_e} q_{xE}^{t} \\ & - \frac{\Delta x \Delta y \tau_q}{\alpha \Delta t} q_{xP}^{t-\Delta t} + (\Delta t + \tau_T)\left(q_{y_{ne}}^{t+\Delta t} - q_{y_{se}}^{t+\Delta t} - q_{y_{nw}}^{t+\Delta t} + q_{y_{sw}}^{t+\Delta t}\right) \\ & - \tau_T\left(q_{y_{ne}}^{t} - q_{y_{se}}^{t} - q_{y_{nw}}^{t} + q_{y_{sw}}^{t}\right) \end{aligned}$$

Here, the symbols $t+\Delta t$, t, and $t-\Delta t$ represent the current time step, previous time step, and previous to previous time step, respectively.

Similarly, the discretized form of the equation for the heat flux component in the y-direction of Eq. (11) can be obtained.

Once the heat flux component in both directions has been obtained, the Eq. (10) has been discretized using the FVM to determine the temperature distribution at the current time instant using Eq. (23) [21].

$$T_P^{t+\Delta t} = T_P^{t} + \frac{\Delta t}{\rho c_v \Delta x}\left(q_{xw}^{t+\Delta t} - q_{xe}^{t+\Delta t}\right) + \frac{\Delta t}{\rho c_v \Delta y}\left(q_{y_s}^{t+\Delta t} - q_{y_n}^{t+\Delta t}\right) \tag{23}$$

The obtained numerical results have been compared with the analytical solution. The analytical solution of the DPL model-based bio-heat transfer equation has been obtained using the FIT technique, which is discussed in Section 4.

4. Analytical modeling

In this section, the analytical solution of the DPL model-based bio-heat transfer equation is obtained using the FIT technique. Then, the algorithm for coupling this analytical solution with a solution of RTE is discussed to determine the laser-irradiated biological tissue.

To determine the temperature distribution inside the laser-irradiated biological tissue, the DPL model-based bio-heat transfer equation in terms of temperature is obtained using Eqs. (4) and (10), which can be expressed as [22]:

$$\begin{aligned} \rho c_v \tau_q \frac{\partial^2 T}{\partial t^2} + \left(\rho c_v + \tau_q \omega_b \rho_b c_b\right)\frac{\partial T}{\partial t} = {} & k\left[\nabla \bullet (\nabla T) + \tau_T \frac{\partial[\nabla \bullet (\nabla T)]}{\partial t}\right] + \omega_b \rho_b c_b (T_b - T) \\ & + \tau_q \omega_b \rho_b c_b \frac{\partial T_b}{\partial t} + Q_m + \tau_q \frac{\partial Q_m}{\partial t} \end{aligned} \tag{24}$$

Under the assumptions of T_b = constant=37 °C, Q_m = constant and θ defined as $T - T_b$, the governing equation (Eq. 24) in the two-dimensional Cartesian coordinate system can be expressed as [22]:

$$\rho c_v \tau_q \frac{\partial^2 \theta}{\partial t^2} + \left(\rho c_v + \tau_q \omega_b \rho_b c_b\right) \frac{\partial \theta}{\partial t} = k \left[\frac{\partial^2}{\partial x^2} \left(\theta + \tau_T \frac{\partial \theta}{\partial t} \right) + \frac{\partial^2}{\partial y^2} \left(\theta + \tau_T \frac{\partial \theta}{\partial t} \right) \right] - \omega_b \rho_b c_b \theta + Q_m \tag{25}$$

Eq. (24) can also be expressed in a cylindrical coordinate system for axisymmetric biological tissue by Eq. (26) [23].

$$\rho c_v \tau_q \frac{\partial^2 \theta}{\partial t^2} + \left(\rho c_v + \tau_q \omega_b \rho_b c_b\right) \frac{\partial \theta}{\partial t} = k \left[\frac{\partial^2 \theta}{\partial r^2} + \frac{1}{r} \frac{\partial \theta}{\partial r} + \frac{\partial^2 \theta}{\partial z^2} + \tau_T \frac{\partial}{\partial t} \left(\frac{\partial^2 \theta}{\partial r^2} + \frac{1}{r} \frac{\partial \theta}{\partial r} + \frac{\partial^2 \theta}{\partial z^2} \right) \right] - \omega_b \rho_b c_b \theta + Q_m \tag{26}$$

The FIT technique [24] converts the partial differential equation into the ordinary differential equation by removing the space derivative using the integral transformation pair. Thus, the ordinary differential equation is only the function of time. This equation with the transformed initial conditions is solved. The transformed temperature can be inverted back into the temperature using the inversion formula.

The boundary conditions Eqs. (27)–(30) and initial conditions Eqs. (31) and (32) required for solving Eq. (25) are given below:

$$\theta(0, y, t) = 0 \tag{27}$$

$$\theta(L, y, t) = 0 \tag{28}$$

$$\theta(x, 0, t) = 0 \tag{29}$$

$$\frac{\partial \theta\left(x, L_y, t\right)}{\partial y} + H\theta\left(x, L_y, t\right) = H\theta_\infty \tag{30}$$

$$\theta(x, y, 0) = 0 \tag{31}$$

$$\frac{\partial \theta(x, y, 0)}{\partial t} = 0 \tag{32}$$

Here $H = \frac{h}{k}$.

The integral transform pair for a given function $\theta(x, y, t)$concerning the x variable can be expressed as below

Inversion formula:

$$\theta(x, y, t) = \sum_{p=1}^{\infty} \frac{X\left(\nu_p, x\right)}{N\left(\nu_p\right)} \bar{\theta}\left(\nu_p, y, t\right) \tag{33}$$

Integral transform:

$$\bar{\theta}\left(\nu_p, y, t\right) = \int_{x'=0}^{L} X\left(\nu_p, x'\right) \theta(x', y, t) dx' \tag{34}$$

Similarly, the integral transform pair for the function $\overline{\theta}(\nu_p, y, t)$ concerning the y variable can be expressed as

Inversion formula:

$$\overline{\theta}(\nu_p, y, t) = \sum_{m=1}^{\infty} \frac{Y(\eta_m, y)}{N(\eta_m)} \tilde{\overline{\theta}}(\nu_p, \eta_m, t) \tag{35}$$

Integral transform:

$$\tilde{\overline{\theta}}(\nu_p, \eta_m, t) = \int_{y'=0}^{W} Y(\eta_m, y')\overline{\theta}(\nu_p, y', t)dy' \tag{36}$$

Here, ν_p,η_m are eigenvalues, $X(\nu_p, x)$, $Y(\eta_m, y)$ are Eigen function and $N(\nu_p)$, $N(\eta_m)$ are the norm. The eigenvalues, Eigen function, and norm for the givens set boundary conditions Eqs. (27)–(30) can be found elsewhere [24].

Multiplying both sides of Eq. (25) by the operator $\int_{x'=0}^{L} X(\nu_p, x')dx'$ and using the Eq. (34) and Eqs. (27) and (28), the derivative concerning x variables is removed from Eq. (25), and the resulting equation is in terms of $\overline{\theta}(\nu_p, y, t)$. Similarly, the derivative concerning y variables is removed from this transformed governing equation by multiplying both sides by $\int_{y'=0}^{W} Y(\eta_m, y')dy'$ and using the Eq. (36) and the transformed boundary conditions of Eqs. (29) and (30), we get [22]

$$\frac{d^2\tilde{\overline{\theta}}(\nu_p, \eta_m, t)}{dt^2} + a\frac{d\tilde{\overline{\theta}}(\nu_p, \eta_m, t)}{dt} + b\tilde{\overline{\theta}}(\nu_p, \eta_m, t) = \frac{1}{\rho c_v \tau_q}\left(kH\theta_\infty \sin\eta_m L_y \frac{[1-(-1)^p]}{\nu_p}\right) \tag{37}$$

where $a = \frac{k\tau_T(\nu_p^2+\eta_m^2)+(\rho c_v+\tau_q\omega_b\rho_b c_b)}{\rho c_v \tau_q}$, $b = \frac{k(\nu_p^2+\eta_m^2)+\omega_b\rho_b c_b}{\rho c_v \tau_q}$

Eq. (37) is the second-order non-homogenous ordinary differential equation, and its solution using the transformed initial conditions has been obtained, which is given below [22]:

$$\theta(x, y, t_{n+1}) = \sum_{p=1}^{\infty} H\theta_\infty \frac{[1-(-1)^p]}{\nu_p} \frac{sinh\, b_1(p)y}{b_1(p)\, cosh\, b_1(p)W + Hsinhhb_1(p)W} \frac{sin\, \nu_p x}{N(\nu_p)}$$

$$+ \sum_{p=1}^{\infty}\sum_{m=1}^{\infty}\left[\tilde{\overline{\theta}}(\nu_p,\ \eta_m,\ t_n) - \left(\frac{kH\theta_\infty \sin\eta_m W}{\rho c_v \tau_q b}\right)\left[\frac{1-(-1)^p}{\nu_p}\right]\right]\left[\left(\frac{\gamma+\omega_1}{2\omega_1}\right) exp\{-(\gamma-\omega_1)\Delta t\}\right.$$

$$\left. -\left(\frac{\gamma-\omega_1}{2\omega_1}\right) exp\{-(\gamma+\omega_1)\Delta t\}\right] \frac{sin\, \nu_p x}{N(\nu_p)} \frac{sin\, \eta_m y}{N(\eta_m)}$$

$$+ \sum_{p=1}^{\infty}\sum_{m=1}^{\infty} \frac{d\tilde{\overline{\theta}}(\nu_p, \eta_m, t_n)}{dt}\left[\frac{exp\{-(\gamma-\omega_1)\Delta t\} - exp\{-(\gamma+\omega_1)\Delta t\}}{2\omega_1}\right] \frac{sin\, \nu_p x}{N(\nu_p)} \frac{sin\, \eta_m y}{N(\eta_m)},$$

$$\left(\frac{a}{2}\right)^2 > b \tag{38}$$

$$\theta(x, y, t_{n+1}) = \sum_{p=1}^{\infty} H\theta_{\infty} \frac{[1-(-1)^p]}{\nu_p} \frac{sinh\, b_1(p)y}{b_1(p)\, cosh\, b_1(p)W + Hsinhhb_1(p)W} \frac{sin\, \nu_p x}{N(\nu_p)}$$

$$+ \sum_{p=1}^{\infty}\sum_{m=1}^{\infty}\left[\tilde{\overline{\theta}}(\nu_p, \eta_m, t_n) - \left(\frac{kH\theta_{\infty} \sin \eta_m W}{\rho c_v \tau_q b}\right)\left[\frac{1-(-1)^p}{\nu_p}\right]\right]\left[\frac{exp\,(-\gamma\Delta t)}{\omega_2}\right][\omega_2 \cos \omega_2 \Delta t$$

$$+\gamma \sin \omega_2 \Delta t] \frac{sin\, \nu_p x}{N(\nu_p)} \frac{sin\, \eta_m y}{N(\eta_m)} + \sum_{p=1}^{\infty}\sum_{m=1}^{\infty} \frac{d\tilde{\overline{\theta}}(\nu_p, \eta_m, t_n)}{dt}\left[\frac{\sin \omega_2 \Delta t exp(-\gamma \Delta t)}{\omega_2}\right] \frac{sin\, \nu_p x}{N(\nu_p)} \frac{sin\, \eta_m y}{N(\eta_m)},$$

$$\left(\frac{a}{2}\right)^2 < b \tag{39}$$

where $\omega_1 = \sqrt{\left(\frac{a}{2}\right)^2 - b}$, and $\omega_2 = \sqrt{b - \left(\frac{a}{2}\right)^2}$

It is to be noted here that the subscript "n" denotes the previous time instant while subscript "n+1" represents the current time instant. Here, $\Delta t = t_{n+1} - t_n$ and $\tilde{\overline{\theta}}(\nu_p, \eta_m, t_n)$is the temperature distribution at the previous time instant t_n. If $t_n = 0$, Eqs. (31) and (32) is used as the initial conditions.

When multiplying both sides of Eq. (37) by τ_q and then substituting $\tau_q = \tau_T = 0$ into Eq. (37), the DPL model-based bio-heat transfer equation becomes the Fourier model-based bio-heat transfer equation, and its solution is expressed by Eq. (40).

$$\theta(x, y, t_{n+1}) = \sum_{p=1}^{\infty} H\theta_{\infty} \frac{[1-(-1)^p]}{\nu_p} \frac{sinh\, b_1(p)y}{b_1(p)\, cosh\, b_1(p)W + Hsinhhb_1(p)W} \frac{sin\, \nu_p x}{N(\nu_p)}$$

$$+ \sum_{p=1}^{\infty}\sum_{m=1}^{\infty}\left[\tilde{\overline{\theta}}(\nu_p, \eta_m, t_n)\right.$$

$$\left. - \left(\frac{kH\theta_{\infty} \sin \eta_m W}{\rho c_v b}\right)\left[\frac{1-(-1)^p}{\nu_p}\right]\right] exp\,(-b\Delta t) \frac{sin\, \nu_p x}{N(\nu_p)} \frac{sin\, \eta_m y}{N(\eta_m)} \tag{40}$$

Following the approach discussed in Section 2, the temperature rise due to a single pulse (ΔT) was first obtained by solving Eq. (9). After that, this temperature rise has been added to the temperature distribution already obtained for the previous time instant to determine the temperature distribution at the current time instant:

$$\tilde{\overline{\theta}}(\nu_p, \eta_m, t_n) = \tilde{\overline{\theta}}(\nu_p, \eta_m, t_n) + \Delta\theta(\nu_p, \eta_m) \tag{41}$$

where

$$\Delta\theta(\nu_p, \eta_m) = \int_{x'=0}^{L}\int_{y'=0}^{W} \Delta T(x', y') \sin \nu_p x' \sin \eta_m y' dx' dy' \tag{42}$$

The trapezoidal rule has been employed for evaluating the double integral appearing on the right-hand side of Eq. (42).

Similarly, the analytical solution of the DPL model-based bio-heat transfer equation in the cylindrical coordinate system Eq. (26) can be obtained using FIT, and the details can be given in elsewhere [23].

5. Results and discussion

The results obtained using the solutions presented in Sections 3 and 4 have been discussed in this section. First, the spatial variation of the two components of the total intensity (collimate and diffused) was calculated at the location of the point of laser irradiation (Location 1 having coordinates of x = 0.98 mm and y = 2 mm) and the center of the domain (Location 2 having coordinates of x = 0.98 mm and y = 0.98 mm) to understand the laser-tissue interaction. After that, the temperature rise due to single-pulse and train of pulses was calculated at the same two locations. Then, the comparative study between the Fourier and non-Fourier (C-V and DPL) heat conduction model were carried out. Furthermore, numerical results were compared with the analytical results obtained using the FIT.

When light propagates through the biological tissue, the intensity has two components: collimated and diffused. The collimated component of the intensity represents the variation of the ballistic photons, which don't undergo any multiple scattering events, and its spatial variation at different time instants is shown in **Figure 2a**. The figure shows that the collimated intensity is maximum at y = 2 mm because it is the point of laser irradiation (Location 1). As expected, the collimated intensity decreases exponentially following the Beer-Lambert law. The collimated intensity is only available when time is less than or equal to the pulse width; otherwise, the value of the collimated intensity is zero.

Variation of the diffuse component of light intensity with respect to the depth (y) of the tissue phantom at various time instants is shown in **Figure 2b**. At the initial time instant (t = 4.6667 ps), the magnitude of diffuse light is significantly much smaller than the collimated intensity. Furthermore, in contrast to the profile seen for the collimated component wherein the maxima were observed at the top surface of the tissue phantom (the point of laser irradiation), the maxima of the diffuse

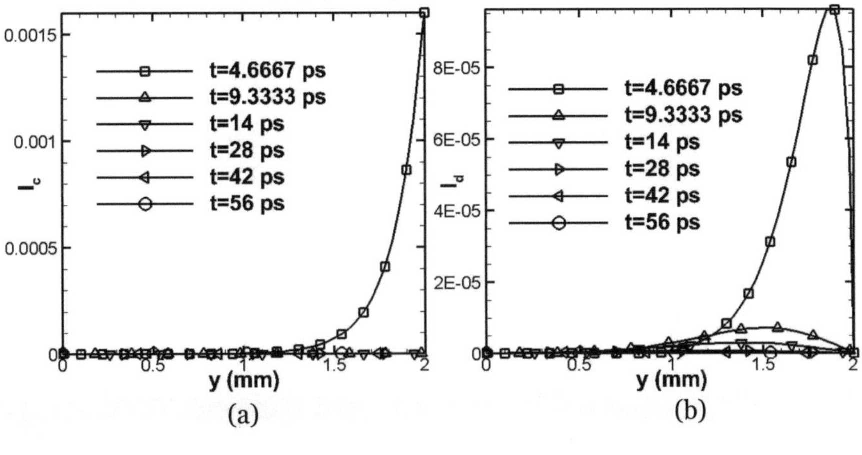

Figure 2.
Spatial variation of collimated (a) and diffuse (b) components of light along a section passing through x = 0.98 *mm [3].*

component are observed at points below the top surface inside the body of the tissue phantom. This trend can be attributed to the scattering properties of the tissue domain, which affect the propagation of the diffuse component of the light intensity.

Figure 3a shows the temporal variation of the temperature at the same two locations considered in the biological tissue subjected to single pulse laser irradiation. As expected, the change in temperature at Location 1 (the point of laser irradiation) is more significant in comparison to Location 2. It is also to be seen from the figure that the rate of temperature rise is quite rapid during the first 5 ps, which is almost equal to the width of the laser pulse employed for irradiation since the laser power is available for this complete duration. Furthermore, the temperature rise achieves a constant value after nearly 20 ps. Because of this observation, the local temperature profile at t = 180 ps was used as the initial temperature field for solving the bio-heat transfer equation (Eq. 10) for determining the temperature distribution within biological tissue subjected to a train of laser pulses.

The Pennes bio-heat transfer equation was solved to determine the temperature distribution inside the biological tissue subjected to the train of pulses (repetition rate: 1 kHz). **Figure 3b** shows the temporal variation of temperature at the same two locations. The figure shows that the temperatures at these locations increase during the first one second, and thereafter a decay in the temperature values can be seen. It is expected since the total duration of laser irradiation is only one second. The temperature rise is comparatively much higher at Location 1, where the laser pulse strikes the sample, than in the rest of the region within the biological tissue. The temperature drop after one second is also steeper at the top wall of the tissue phantom as it is directly exposed to the ambient conditions.

The efficacy of laser-based photo-thermal therapy for selective destruction of cancerous cells depends on accurately predicting the temperature distribution inside the laser-irradiated biological tissue. Because of this, various heat conduction models, e.g., Fourier, C-V, DPL, etc., have been developed by researchers in the past. The need for the non-Fourier heat conduction model is realized because the photons travel with a finite speed through the biological tissues, contrary to the infinite speed of thermal wave propagation assumed in the Fourier model. So, we compared the temperature

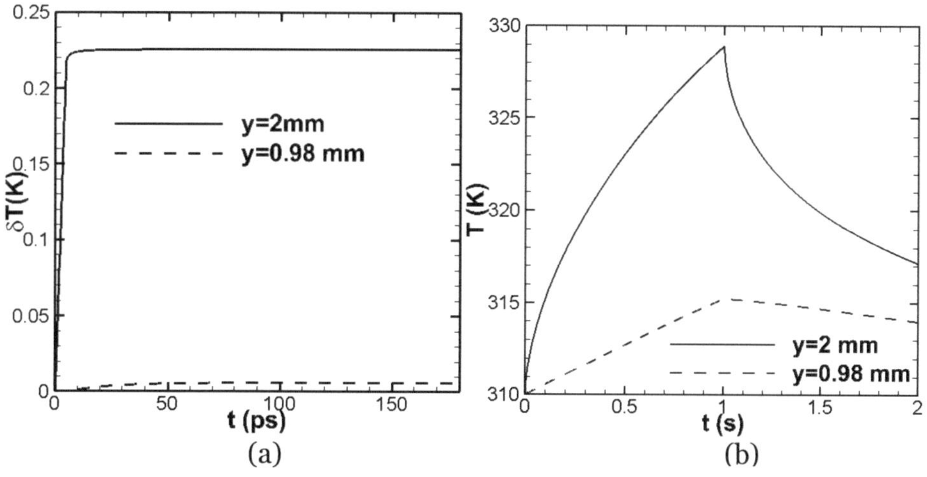

Figure 3.
(a) Temporal profile of temperature rise in tissue subjected to single pulse laser irradiation at x = *0.98 mm; (b) Temporal profile of temperature subjected to a train of pulse at* x = *0.98 mm [3].*

distribution obtained using the Fourier and non-Fourier heat conduction models to study the thermal response of laser-irradiated biological tissue phantom. **Figure 4** shows the two-dimensional temperature distribution inside the laser-irradiated biological tissue at different time instants for Fourier and non-Fourier (C-V and DPL) models. While a nearly uniform diffusion of heat is to be seen in the contour plots corresponding to the Fourier model, a distinct wave nature in the thermal profiles is associated with the predictions of non-Fourier (C-V and DPL) heat conduction models. The oscillations in temperature values penetrate within the body of the tissue phantom as time progresses. The C-V heat conduction model predicts the maximum temperature values at any given time instant, followed by the DPL-based model. The C-V model also results in maximum amplitude of oscillations seen in the thermal wavefront compared to the DPL model.

The numerical results obtained using the FVM were compared with analytical results obtained using the FIT. **Figure 5** shows the analytical and numerical results for different heat conduction models (Fourier and non-Fourier). The temporal profiles of temperature without markers represent the results based on the FIT-based analytical solution, and those with markers represent the numerical results. The figure shows that the results obtained using numerical simulations predict relatively lower values of temperatures at Locations 1 and 2 in comparison with those obtained based on the

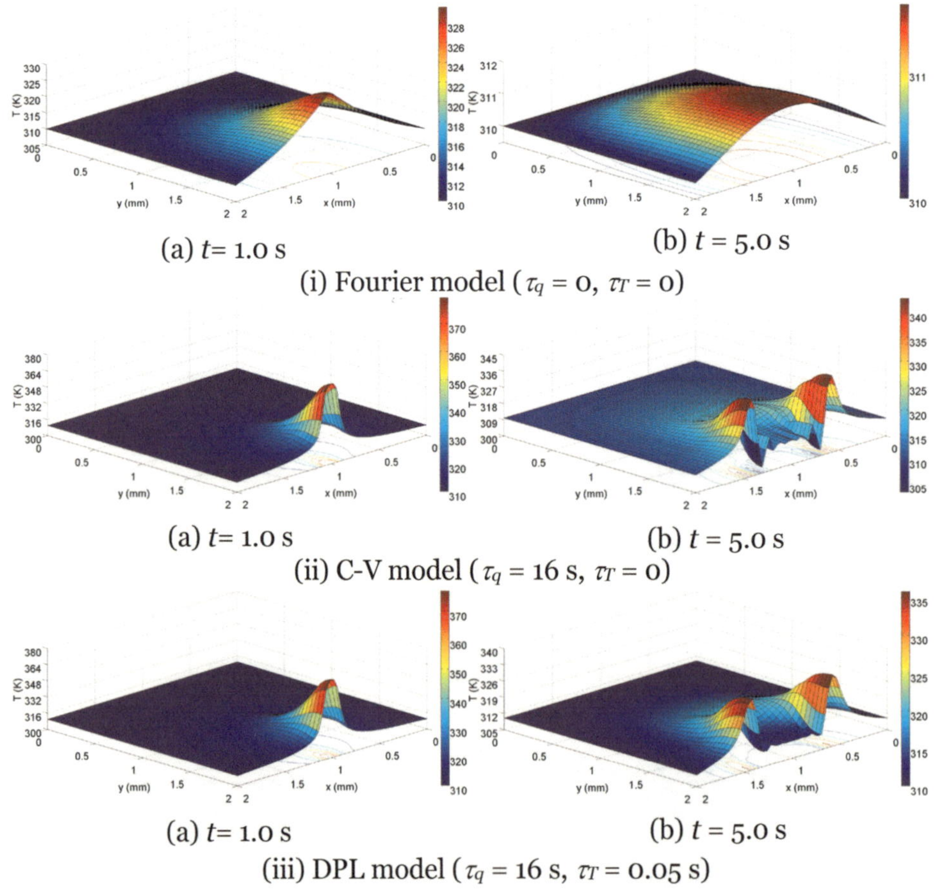

Figure 4.
Two-dimensional temperature distribution at different time instants (i) Fourier (ii) C-V (iii) DPL model [21].

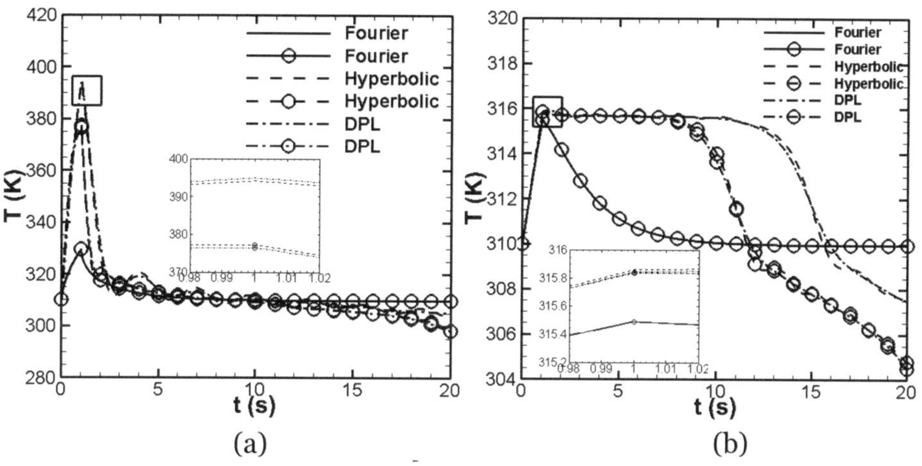

Figure 5.
Comparison of temporal profiles of temperature predicted using Fourier and non-Fourier heat conduction models at Location 1 (a) and 2 (b) (Without marker: Analytical results; With marker: numerical results) [22].

analytical approach. It is also to be seen from the figure that the Fourier heat conduction model predicts relatively lower temperature values than those calculated using non-Fourier conduction models, which is to be attributed to the infinite speed of propagation of thermal waves considered in the conventional Fourier model.

Furthermore, the C-V heat conduction model predicts the highest temperature values (shown in the inset of **Figure 5a** and **b** for better clarity), while the DPL model predicts the temperatures that lie in between Fourier and C-V heat conduction models. The observed trend may be explained based on the C-V heat conduction model only considers the effect of the thermal relaxation time associated with heat flux (τ_q). In contrast, the dual-phase lag model considers the coupled effects of non-zero values of relaxation times related to the temperature gradient (τ_T) and heat flux (τ_q). In physical terms, the phase lag associated with the temperature gradients, i.e., τ_T tends to suppress the amplitude of the thermal wavefront. Because the value of τ_T is zero in the C-V heat conduction model, the resultant temperature profiles are relatively free of such dampening effects. Hence, the absolute values of temperatures predicted based on this form of non-Fourier heat conduction model are expected to be higher than predicted using the DPL model, wherein these two relaxation times are considered non-zero. The profiles shown in **Figure 5** support this observation wherein the magnitude of temperatures as predicted using the DPL model lies between those obtained using the C-V heat conduction model (maximum values) and the Fourier-based heat conduction model (minimum values).

Figure 5b shows the temporal variation of temperature at Location 2. The profiles corresponding to the Fourier heat conduction model show a sudden drop in temperature values immediately after the laser power is switched off ($t > 1.0$ s). On the other hand, because of the phase lag terms associated with temperature gradients and/or heat flux, the temperature profiles predicted using the non-Fourier heat conduction models (hyperbolic and DPL) show nearly constant values of temperature for a relatively long period of time ($t \leq 11$ s) before the drop in temperature values starts. It indicates the prolonged (sustained) effects of thermal energy deposited at a given spatial location within the body of the tissue phantom, according to non-Fourier heat conduction models.

6. Conclusion

The present chapter discusses the modeling of the laser-irradiated biological tissue to understand its thermal behavior, which may help improve the efficacy of the laser-based photo-thermal therapy to destroy the cancerous cells with minimal damage to the surrounding healthy tissue. The light propagation through the biological tissue was mathematically modeled using the RTE. The RTE was solved using the DOM to determine the intensity distribution inside the biological tissue subjected to short-pulse laser irradiation. Once the intensity distribution was obtained, the divergence of radiative heat flux was calculated, which acts as the source term in the Fourier/non-Fourier model-based bio-heat transfer equation to determine the temperature distribution inside the laser-irradiated biological tissue. However, it is a multi-time scale problem because of the two different time scales used to solve the RTE and bio-heat transfer equation. So, the algorithm for solving this multi-time scale was presented. The non-Fourier model-based bio-heat transfer equation was numerically solved using the FVM to determine the temperature distribution. The numerical results were compared with the analytical results obtained using the FIT technique and found that the numerical solution predicted relatively lower temperature values than the analytical solution. A comparative study between the Fourier and non-Fourier (C-V and DPL) models was conducted and found that the temperature predicted using the DPL model lies between the C-V and Fourier models.

Acknowledgements

The author sincerely acknowledges the technical input provided by Prof. Atul Srivastava, IIT Bombay, India, during the preparation of this book chapter. The constant support and encouragement received from him over the last several years are gratefully acknowledged. The author also acknowledges the continuous support received from IIT Bombay and NIT Rourkela during the preparation of this work.

Author details

Sumit Kumar
Department of Mechanical Engineering, National Institute of Technology Rourkela, India

*Address all correspondence to: kumarsumit@nitrkl.ac.in

References

[1] National Cancer Institute. What is Cancer. 2021. Available from: https://www.cancer.gov/about-cancer/understanding/what-is-cancer

[2] World Health Organization. Cancer. 2022. Available from: https://www.who.int/news-room/fact-sheets/detail/cancer

[3] Kumar S, Srivastava A. Numerical investigation of thermal response of laser irradiated tissue phantoms embedded with optical inhomogeneities. International Journal of Heat and Mass Transfer. 2014;77:262-277

[4] Jaunich M, Raje S, Kim K, Mitra K, Guo Z. Bio-heat transfer analysis during short pulse laser irradiation of tissues. International Journal of Heat and Mass Transfer. 2008;**51**:5511-5521. DOI: 10.1016/j.ijheatmasstransfer.2008.04.033

[5] Litvinova KS, Rafailov IE, Dunaev AV, Sokolovski SG, Rafailov EU. Non-invasive biomedical research and diagnostics enabled by innovative compact lasers. Progress in Quantum Electronics. 2017;**56**:1-14. DOI: 10.1016/j.pquantelec.2017.10.001

[6] Zhou J, Zhang Y, Chen JK. An axisymmetric dual-phase-lag bioheat model for laser heating of living tissues. International Journal of Thermal Sciences. 2009;**48**:1477-1485. DOI: 10.1016/j.ijthermalsci.2008.12.012

[7] Fanjul-Vélez F, Romanov OG, Arce-Diego JL. Efficient 3D numerical approach for temperature prediction in laser irradiated biological tissues. Computers in Biology and Medicine. 2009;**39**:810-817. DOI: 10.1016/j.compbiomed.2009.06.009

[8] Guo C, Wan Z, Kim SK, Kosaraju K. Comparing diffusion approximation with radiation transfer analysis for light transport in tissues. Optical Review. 2003;**10**:415-421. DOI: 10.1007/s10043-003-0415-y

[9] Zhang R, Verkruysse W, Aguilar G, Nelson JS. Comparison of diffusion approximation and Monte Carlo based finite element models for simulating thermal responses to laser irradiation in discrete vessels. Physics in Medicine and Biology. 2005;**50**:4075-4086. DOI: 10.1088/0031-9155/50/17/011

[10] Modest MF. Radiative Heat Transfer. 3rd ed. Oxford: Academic Press; 2013

[11] Rath P, Mishra SC, Mahanta P, Saha UK, Mitra K. Discrete transfer method applied to transient radiative transfer problems in participating medium. Numerical Heat Transfer, Part A. 2003;**44**:183-197. DOI: 10.1080/713838195

[12] Chai JC, Hsu P, Lam YC. Three-dimensional transient radiative transfer modeling using the finite-volume method. Journal of Quantitative Spectroscopy and Radiative Transfer. 2004;**86**:299-313. DOI: 10.1016/j.jqsrt.2003.08.008

[13] Guo Z, Kumar S. Three-dimensional discrete ordinates method in transient radiative transfer. Journal of Thermophysics and Heat Transfer. 2002;**16**:289-296. DOI: 10.2514/2.6689

[14] Mishra SC, Chugh P, Kumar P, Mitra K. Development and comparison of the DTM, the DOM and the FVM formulations for the short-pulse laser transport through a participating medium. International Journal of Heat

and Mass Transfer. 2006;**49**:1820-1832. DOI: 10.1016/j.ijheatmasstransfer. 2005.10.043

[15] Mitra K, Kumar S. Development and comparison of models for light-pulse transport through scattering–absorbing media. Applied Optics. 1999;**38**:188-196. DOI: 10.1364/AO.38.000188

[16] Hunter B, Guo Z. Comparison of the discrete-ordinates method and the finite-volume method for steady-state and ultrafast radiative transfer analysis in cylindrical coordinates. Numerical Heat Transfer, Part B. 2011;**59**:339-359. DOI: 10.1080/10407790.2011.572719

[17] Dombrovsky LA, Timchenko V, Jackson M. Indirect heating strategy of laser-induced hyperthermia: An advanced thermal model. International Journal of Heat and Mass Transfer. 2012; **55**:4688-4700. DOI: 10.1016/j. ijheatmasstransfer.2012.04.029

[18] Kumar S, Srivastava A. Numerical investigation of the influence of pulsatile blood flow on temperature distribution within the body of laser-irradiated biological tissue phantoms. International Journal of Heat and Mass Transfer. 2016; **95**:662-677. DOI: 10.1016/j. ijheatmasstransfer.2015.12.023

[19] Mitra K, Kumar S, Vedavaraz A, Moallemi M. Experimental evidence of hyperbolic heat conduction in processed meat. ASME Journal of Heat Transfer. 1995;**117**:568-573. DOI: 10.1115/1.2822615

[20] Kim K, Guo Z. Multi-time-scale heat transfer modeling of turbid tissues exposed to short-pulsed irradiations. Computer Methods and Programs in Biomedicine. 2007;**86**:112-123. DOI: 10.1016/j.cmpb.2007.01.009

[21] Kumar S, Srivastava A. Thermal analysis of laser-irradiated tissue phantoms using dual phase lag model coupled with transient radiative transfer equation. International Journal of Heat and Mass Transfer. 2015;**90**:466-479. DOI: 10.1016/j.ijheatmasstransfer. 2015.06.077

[22] Kumar S, Srivastava A. Finite integral transform-based analytical solutions of dual phase lag bio-heat transfer equation. Applied Mathematical Modelling. 2017;**52**:378-403

[23] Kishore P, Kumar S. Analytical investigation of non-Fourier bioheat transfer in the axisymmetric living tissue exposed to pulsed laser heating using finite integral transform technique. Transactions of the ASME: Journal of Heat Transfer. 2021;**143**:121-201

[24] Hahn DW, Ozisik MN. Heat Conduction. New Jersey: John Wiley & Sons, Inc.; 2012

Chapter 4

Femtosecond Laser Micro-/ Nano-Texturing to Die Substrates for Fine Imprinting to Products

Tatsuhiko Aizawa, Tadahiko Inohara, Yohei Suzuki and Tomomi Shiratori

Abstract

A femtosecond laser micro–/nano-texturing was proposed to fabricate the coated and surface treated dies with the tailored textures for surface decoration and surface property control of metal, polymer and glass products. The polygonal model for microtextures with nanotextures by the LIPSS-effect was utilized to fabricate a DLC-coated SKD11 die with a star-shaped emblem. This die was set up into the cassette die set for directly imprinting this emblem into aluminum alloy and PET sheets. The periodic surface structure was synthesized as a surface geometry model to build up the super-hydrophobic surface on the nitrogen supersaturated AISI316 die. This die was also set up into a hot stamping system to directly imprint the hydrophobic surface onto the phosphorous glass products. Through the femtosecond laser micro–/nano-texturing and CNC-imprinting, the metal, polymer and glass product surfaces were optically decorated to have color grating and plasmonic brilliance and functionally controlled to be hydrophobic.

Keywords: femtosecond laser micro–/nano-texturing, surface decoration, surface property control, DLC coating die, nitrogen supersaturated die, imprinting, color grating, surface plasmonic brilliance, hydrophobicity

1. Introduction

Surface decoration is an essential method for copy-proof of originally designed products. The holograph and color-grating techniques are utilized in a newly designed 10,000-yen paper billets [1], where a few holographic symbols are imprinted onto this billet to be free from forgery together with the accurately printed portrait of the late Mr. E. Shibusawa, a famous founder of enterprises in Japan. The color-grating method with surface plasmonic design is also utilized to decorate the polymer surface [2] and to modify the original surface properties [3]. In the conventional approach, the designed micro–/nano-textures are printed onto a plastic foil, which is further pasted onto the product surface, as depicted in **Figure 1a**. This approach is easy to be done but it has always a risk of foil delamination from the product surface in daily usage. **Figure 1b** depicts the two-step procedure where the designed textures are cut

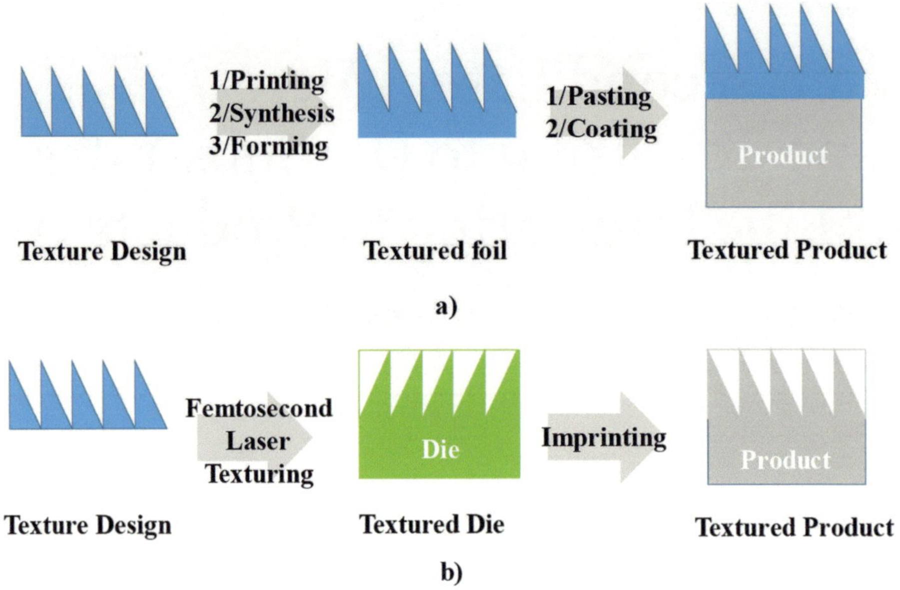

Figure 1.
Two approaches to form the micro–/nano-textures onto the product surface. a) Indirect formation of textures by pasting the textured foils onto the products, and b) direct imprinting of textures onto the products.

into the die surface to transcribe this negative pattern to the product surface [4–6]. The femtosecond laser micro–/nano-texturing has been highlighted as a flexible tool to form the tailored surface and interface profiles onto any material substrates [4, 5]. Hard coating layers such as DLC (Diamond-Like Carbon) and diamond, were micro–/nano-textured by using the femtosecond machining [6, 7]. In particular, as stated in [8, 9], those hard-coated and surface-treated substrates were suitable as a special tool for directly imprinting the shaped micro–/nano-textures onto the work materials.

In this direct imprinting, by using the laser-treated dies, various textures in the order from mm down to sub-μm were transcribed onto the product surfaces. The accurately aligned micro-textures for diffraction optical elements were first shaped onto a die surface by femtosecond laser machining and then imprinted onto the plastic and glass products to fabricate an optical lens with DOE. **Figure 2a** and **b** illustrate the Fresnel-patterned flat lens cross-section and the top view of heat-transferring

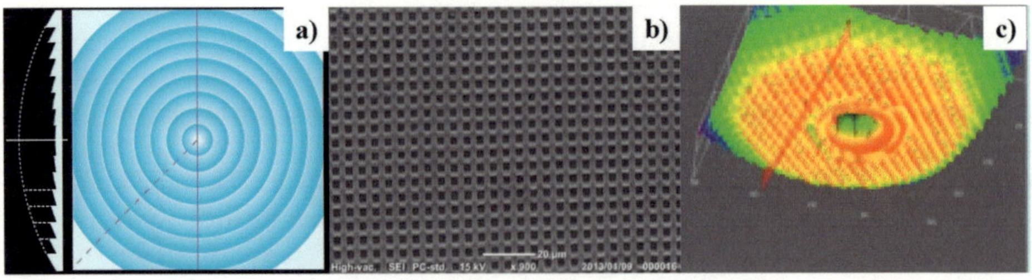

Figure 2.
Various products with directly imprinted micro–/nano-textures. a) Fresnel-patterned flat lens, b) heat-transferring aluminum device with regular micro-cavity alignment, and c) dispensing stainless steel nozzle with micro–/nano-textures around its outlet.

aluminum device with regular micro-cavity array, respectively [10, 11]. Due to the fine micro-cavity alignment with the unit size of 3.5 μm × 3.5 μm, the heat flux in the boiling curve was increased five times higher than the non-textured aluminum plate. **Figure 2c** depicts the laser-treated stainless steel nozzle to dispense sub-nL to pL droplets for inkjet printing and line-drawing [12]. Due to the micro–/nano-textured around the nozzle outlet, the diameter of the dispensed droplet was preserved to be nearly equal to the inner diameter of the nozzle outlet.

In the direct imprinting in **Figure 1**, various kids of die material are selected to each application. Consider the flow stress of product materials in practice. Most of metallic products have a yield stress, the above which they begin to deform elasto-plastically and to shape themselves under the constraint of die surfaces [13]. The oxide glasses are fragile below their glass transition temperature; they are able to be elasto-viscously formed into the tailored optical element shape under the constraint of mold surfaces at the elevated temperature [14]. The die and mold materials are optimally designed to have sufficient hardness against the high flow stress of works, to have high erosion and corrosion toughness in contact to work materials in cold and hot, and to have high chemical stability for high laser-machinability. There are two die and mold design approaches for directly imprinting the tailored micro–/nano-textures into the die and mold substrates by using the femtosecond laser machining [15].

Figure 3a depicts the hard coating die material with significant film thickness. As discussed in [15], CVD (Chemical Vapor Deposition) coated DLC, diamond and β-SiC films have sufficient thickness to be working as a die substrate for laser micro-texturing. On the other hand, the plasma nitriding and carburizing at low temperature are available to make nitrogen and carbon supersaturation into the Fe-Cr alloys and stainless steels, as depicted in **Figure 3b** [9, 16–18]. These supersaturation process provides the nitrogen- and carbon-alloyed layers without nitride and carbide precipitates; those secondary phases often play as an origin of fatigue cracking and deteriorate the original corrosion toughness of chromium-base die substrates. Both coated and surface-treated layers are expected to be working as a die and mold substrate for accurate micro–/nano-texturing with well-defined abrasion behavior.

In the present study, a thick DLC coating is utilized as a die to imprint the laser-textured surface into the metal and polymer sheets by using the CNC (Computer Numerical Control) cold stamping system. The optical properties of textured DLC-die are transcribed onto these sheet surfaces together with geometric and topological coining of die surface profile. This color-grating and plasmonic brilliance of imprinted micro–/nano-textures onto sheets decorate their surface. A thick nitrogen-supersaturated layer is also used as a mold to imprint the laser-textured

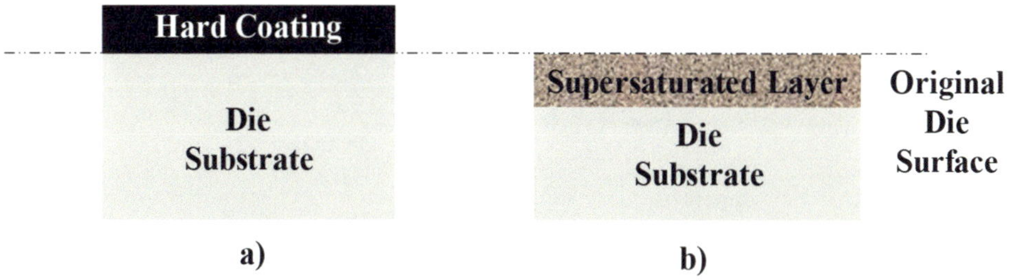

Figure 3.
Two die and mold designs for directly imprinting the tailored micro–/nano-textures into the die and mold substrates by using the femtosecond laser machining.

surface to the phosphorous glass works by using the CNC hot stamping system. The super-hydrophobic die surface is imprinted to the hydrophobic work surface with high contact angle. The glass work surface is controlled from hydrophilic to hydrophobic state by this CNC-imprinting. Owing to the appropriate die and mold material selection, these imprinting processes are free from galling or adhesion wear of work material debris particles. Various applications are discussed to make full use of this surface decoration and surface property control by the femtosecond laser micro–/nano-texturing.

2. Micro-/nano-texturing procedure

Various micro–/nano-textures are required for surface decoration and surface property control to improve the product quality and function. First, a texture design is proposed for surface profiling to be installed by the femtosecond laser texturing. Two types of die substrate materials are utilized for this laser texturing; e.g., thick DLC coated SKD11 die and nitrogen supersaturated AISI316 die. The former die is selected for surface decoration by laser texturing. The latter die is used for surface property control. These dies are utilized to transcribe the die textures into the work materials by the CNC-stamping system. SEM and three-dimensional surface profilometer are employed for characterization of the surface profiles of dies and works.

Texture Design for Surface Decoration and Surface Property Control. In the laser texturing for surface decoration, the symbols, the fonts, the image, the pictures, the patterns, and the figures are represented by the simple geometric model to reduce the efforts to prepare an amount of CAM (Computer Aided Machining) data for laser machining. As shown in **Figure 4a**, each unit geometry is modeled by a polygonal segment, which consists of the lines and dots for laser microtexturing. Nanotextures are induced onto the edges and terraces of micro-textured zones by the LIPSS (Laser-Induced Periodic Surface Structuring)-effect [19]. **Figure 4b** depicts a typical one-dimensional surface profile; some periodic profiles are tailored and synthesized to this profile in order that a Fractal dimension along the laser-scanned direction is optimally controlled for well-defined surface property [20]. In this surface property control, the nanotextures by the LIPSS-effect are formed onto the micro-textured edges and terraces to preserve the self-similar surface with the same Fractal dimension as specified by micro-texturing.

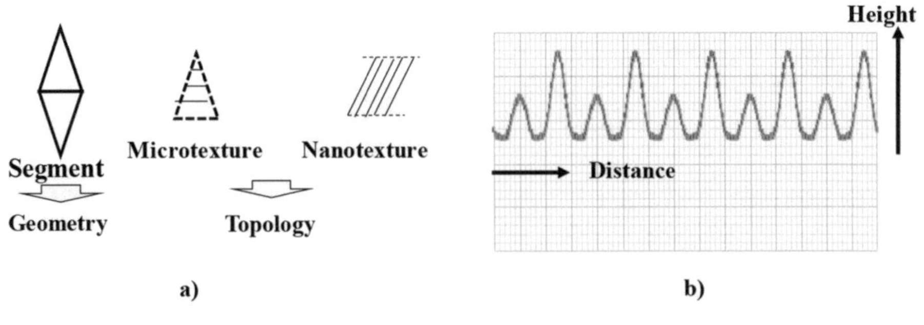

Figure 4.
Texture design for surface decoration and surface property control. a) Polygonal model for laser path control to print the micro-textures, and b) synthesizing the periodic structures into a surface profile.

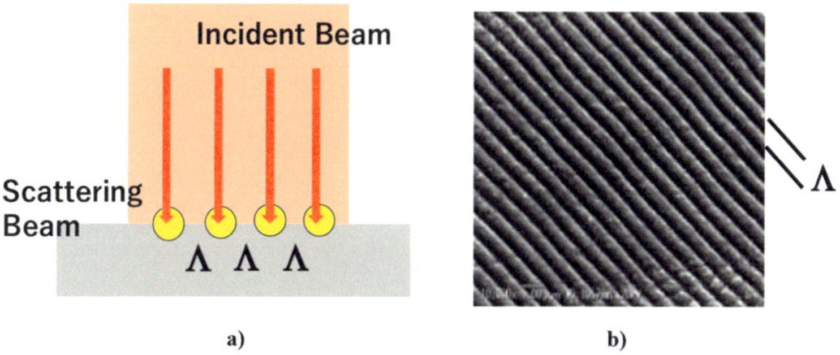

Figure 5.
Simultaneous nanotexturing by LIPSS with microtexturing. a) Schematic view on LIPSS, and b) formation of nanotextures with the LIPSS-period.

As illustrated in **Figure 5**, these LIPSS-ripples are induced by optical interaction between the incident laser beam and the scattered beam by the surface roughness. After [21], this LIPSS-period is affected by the wavelength and fluence in the laser irradiation. Both higher and lower frequency nanotextures against the original wavelength are formed onto the irradiated surface. In addition, the orientation of nanotextures is also controllable by using optical polarization or by twisting the laser beam. In general, this laser nanotexturing is rather insensitive to the work material selection; to be discussed later, the ablation steps by laser irradiation might be influenced by the microstructure of materials.

Femtosecond Laser Texturing System. A femtosecond laser system (FEM-1; LPS-Works, Co., Ltd., Tokyo, Japan) was used to print the tailored spatial textures directly onto the DLC coating surface. The wavelength (λ) of the laser was 515 nm, with a pulse width of 200 fs and a pulse repetition rate of 400 kHz. The maximum average power was 40 W, and the maximum pulse energy was 50 μJ. The working area was 300 mm × 300 mm. In practical operation, a working plate with the size of 280 mm × 150 mm was placed on the work table as depicted in **Figure 6**. The irradiation power of a single pulse is estimated to be 0.25 GW. This high-power irradiation in the 200-fs interval drives a well-defined ablation into the targeting materials. The femtosecond laser machining process was controlled by the CAM (Computer Aided Manufacturing) data. In this experiment, each microtexture is represented by the assembly of line segments. Nanotexture is cut into each micro-texture by the LIPSS effects. In this LIPSS, each nano-groove is formed by the nonlinear optical interaction between the controlled incidental laser beam and the traveling beam on the surface. Depending on the laser irradiation parameters and the surface condition, the nano-groove depth (d_L) is uniquely determined; in this case, $d_L \sim 400$ nm. On the other hand, the LIPSS-period (Λ) or the nano-groove width is also determined by the laser processing conditions. In this case, $\Lambda \sim 300$ nm.

Die substrate material selection. Different from conventional metal forming, a die substrate material has an amorphous carbon film or a nano-size grain-structured surface layer. Otherwise, the grain boundaries are easy to be imprinted together with the micro−/nano-textures when using the polycrystalline metals, alloys, ceramics, and thermets with the normal grain sizes as used in the normal die and mold. In the following experiments, both the DLC-coated SKD11 die and the nitrogen supersaturated AISI316 mold are utilized for femtosecond laser micro−/nano-texturing. In this material selection, how to control the pulsed laser ablation becomes a key to efficiently subtract the amorphous carbon and nitrogen supersaturated Fr−Cr (N). As stated in [5], DLC

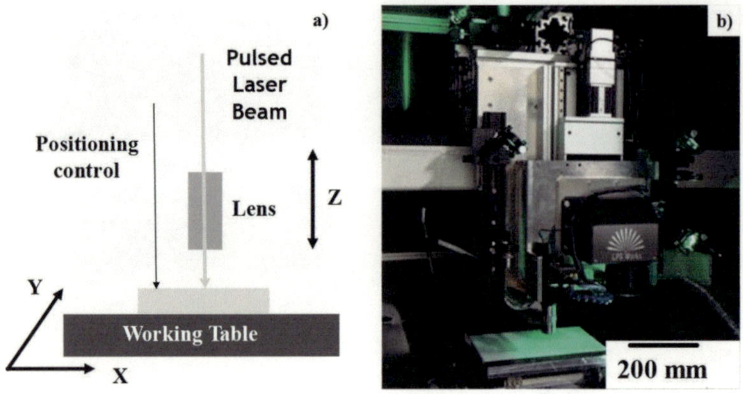

Figure 6.
Femtosecond laser micro−/nano-texturing system. a) A schematic view of laser processing in operation, and b) an overview of the system.

coating is efficiently machined by low-power application without the deposition of carbon particles onto the DLC die. Laser power and fluency must be optimized for laser texturing of nitrogen solute bearing tool steels and stainless steel molds.

CNC-Imprinting. Two types of CNC-stamping systems were utilized to transcribe the original micro−/nano-textures on the dies into the work materials. The cold and warm CNC-stamping system (ZEN90, Hoden-Seimitsu, Co., Ltd.; Kanagawa, Japan) was used for imprinting the textured die surface into the metallic and polymer works with relatively low melting temperature as shown in **Figure 7**. The textured die was placed into the upper die set. Both the upper and lower cassette die-sets were respectively fixed to the upper and lower bolsters of this system, respectively. In the cold and warm imprinting process, the upper bolster was incrementally lowered to imprint the mother textures on the die onto the work surface after the starting position in contact with die surface of the work. The stroke velocity was constant by 0.05 mm/s; various loading schedules can be programmed in this CNC-imprinting system. This cold upsetting process was performed until the total stroke of 150 μm by the applied load of 3 kN [7].

In the hot imprinting system, the IH (Induction Heating)-unit was used for prompt and accurate thermal transient control in **Figure 8**. Both the upper and lower

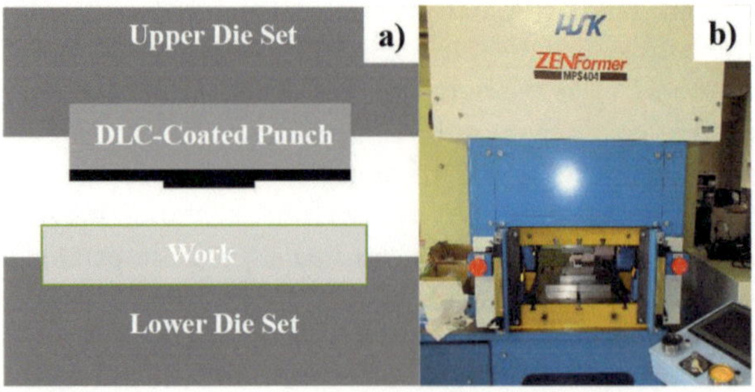

Figure 7.
Cold CNC-stamping system for imprinting the mother textured die onto the metallic and polymer sheets at room temperature. a) A schematic view of cold imprinting process, and b) an overview of cold imprinting system.

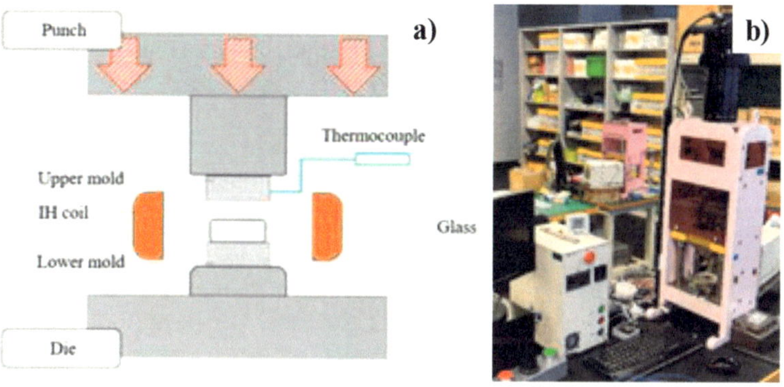

Figure 8.
Hot CNC-stamping system for imprinting the mother textured mold onto the glasses above the glass transition temperature. a) A schematic view of hot imprinting process, and b) an overview of hot imprinting system.

molds were located on the inside of IH-coil for uniform heating. As stated in [22], the heating and cooling steps were PID (Proportional-Integral-Differentiation)-controlled to narrow the temperature deviation of molds within ±1 K in the inline temperature measurement by the embedded thermocouples into the upper mold. Both the loading sequence and temperature history were controlled by the personal computer. How to control the temperature history, is discussed later.

Characterization. SEM (Scanning Electron Microscopy; JOEL, Tokyo, Japan) was utilized for surface analysis on the textured die and work surfaces. Three-dimensional profilometer (NT91001, Bruker AEX Co., Ltd.; Tokyo, Japan) and laser microscopy (Olympus Co., Ltd., Tokyo, Japan) were also used to describe the depth profiles of micro-textures.

3. Femtosecond laser texturing

The femtosecond laser machining system was utilized to make a surface decoration of DLC coating die and to control the surface properties of nitrogen supersaturated die by the laser micro–/nano-texturing.

Die preparation for surface decoration and surface property control. Thick DLC coating was deposited onto the SKD11 substrate with the size of 100 mm × 100 mm × 5 t mm by using the MF (Medium Frequency) – PECVD (Plasma Enhanced Chemical Vapor Deposition) system (KOBELCO, Japan). This DLC-deposited substrate was further cut and finished to a die shape with a head size of 10 mm × 20 mm. **Figure 9a** depicts the DLC-coated SKD11 die. After [23], this amorphous carbon layer has a homogeneous nanostructure with the constant hardness of 22 GPa even by varying the layer thickness and the PECVD processing conditions. The average surface roughness was much less than 0.1 μm. This DLC coating was utilized for laser surface decoration and direct imprinting in cold and warm.

The largest drawback of this DLC coating is low thermal resistance at elevated temperature. As pointed out in [24], the amorphous carbon becomes chemically unstable when the holding temperature is higher than 623 K or 350°C. Hence, another substrate material must be selected for hot imprinting of textures into the glass work materials.

The nitrogen supersaturated AISI420 substrate was utilized as a mold for hot imprinting process. As-machined AISI420, mold with a diameter of 12 mm was

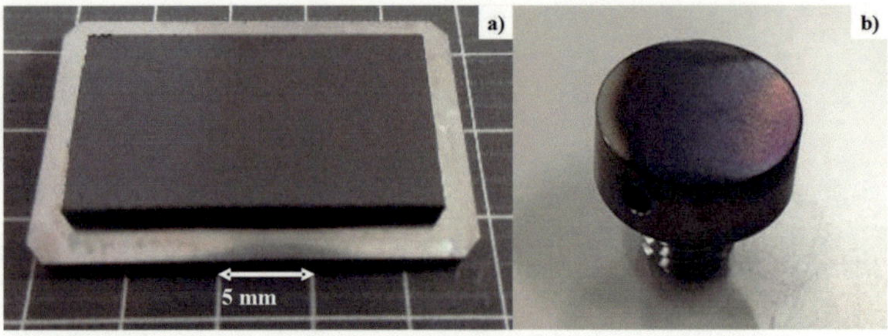

Figure 9.
Two types of dies are to be laser-textured for surface decoration and surface property control. a) DLC coated SKD11 die with the amorphous carbon layer thickness of 20 μm, and b) nitrogen supersaturated AISI420 die with the nitrided layer thickness of 50 μm.

prepared and nitrided at 673 K for 14.4 ks (or 4 h) to increase the surface hardness from the matrix hardness of 260 HV to 1100 HV and to be supersaturated in the 50 μm nitrided layer by higher nitrogen content of 7 mass%. **Figure 9b** depicts the nitrogen supersaturated AISI420 mold. No surface roughing was observed on the top surface of mold; no disturbance in dimension was also detected after nitriding. Without additional grinding and polishing, as nitrided AISI420 mold was employed for laser texturing.

Femtosecond laser micro–/nano-texturing. These dies and molds were used for femtosecond laser micro–/nano-texturing. A star-shaped emblem was textured onto the DLC die to describe the femtosecond laser processing. **Figure 10** depicts this textured emblem, which is represented by eight polygonal segments including the nano-grooves with their tailored orientation. Each segment in this emblem is distinguished by its own color-grating on the micro-textured surface with a pitch of 10 mm. The nanotextured zone was colored by its surface-plasmonic brilliance. In this laser texturing operation, the whole DLC-die surface was once ground down to the depth of 7 μm except for a square area with the size of 6 mm × 6 mm. Each constituent segment was laser-cut and shaped onto this square region to build up an emblem on the DLC die.

The nitrogen supersaturated AISI420 mold was laser-textured to have tailored micro–/nano-grooves with the pitch of 20 μm and the height of 4 μm. **Figure 11** depicts the laser-textured mold surface profile. The average pitch and height of

Figure 10.
DLC coating die with a star-shaped emblem, which was laser micro–/nano-textured.

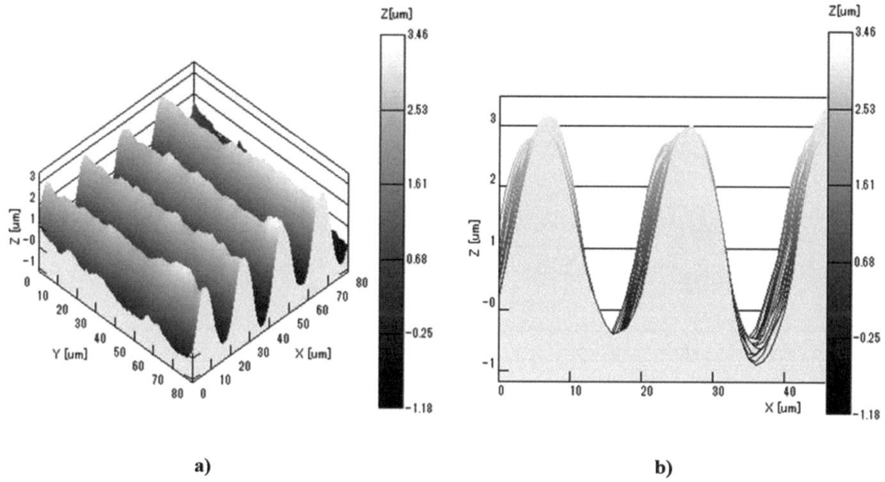

Figure 11.
Surface profiles of the nitrogen supersaturated AISI420 die after laser-texturing along the x-axis with the tailored pitch of 20 μm and the height of 4 μm. a) Bird-view of laser-textured surface, and b) its cross-sectional view.

those micro-grooves were measured to be 19.8 mm and 3.8 mm, respectively. This dimensional accuracy in laser texturing reflects on the surface property. AISI420 mold before texturing was hydrophilic with the contact angle (θ) of 70° while the micro-grooved AISI420 mold became hydrophobic with θ = 140°. After [20, 22, 25, 26], the hydrophobic surface with high contact angle is thought to have high repellency with low falling angle (ϕ). The pure droplet falls along this laser micro-grooved mold surface by ϕ = 25°. This reveals that the textured mold surface in **Figure 11** has sufficiently low surface energy with low wettability.

Characterization. Let us investigate the microstructures of these laser-textured DLC-die and nitrogen supersaturated die surfaces. **Figure 12** depicts the SEM of micro–/nano-textures formed onto the DLC-die with varying magnifications from **Figure 12a–d**. A-zone in **Figure 12a** turns to be **Figure 12b, B**-zone in **Figure 12b**

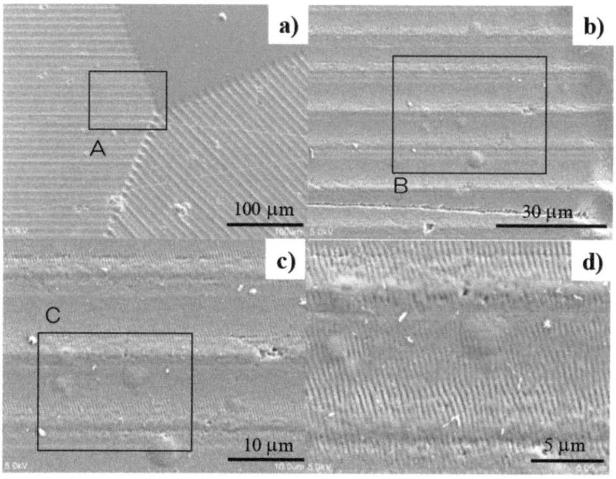

Figure 12.
SEM image on the micro–/nano-textures formed on the DLC coating die with varying the magnitude from left to right.

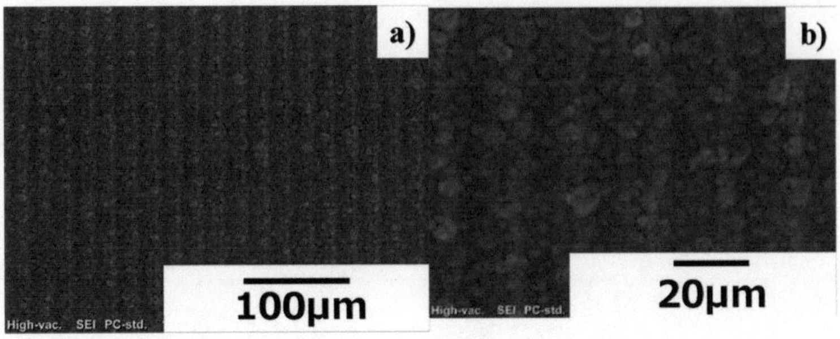

Figure 13.
SEM image on the micro−/nano-textures formed on the nitrogen supersaturated AISI420 mold surface. a) SEM image in low magnification, and b) SEM image in high magnification.

turns to be **Figure 12c**, and **C**-zone in **Figure 12c** turns to be **Figure 12d**. **Figure 12a** and **b** show that each segment consists of a regular alignment of micro-grooves with a pitch of 1-mm. Each micro-groove is formed by two micro-edges and concave terrace. As depicted in **Figure 12c** and **d**, these micro-edges and terraces are nano-textured to have an alignment of ripples with the LIPSS-period of 300 nm. This fine multi-dimensional surface structure characterizes the micro−/nano-textured DLC-die surface.

Figure 13 depicts the SEM image of the textures formed on the nitrided AISI420 mold surface with different magnifications. Micro-edges were regularly formed with a pitch of 20 μm. The nano-grooves were also formed in the longitudinal direction on the terrace between two adjacent micro-edges. This regular alignment of micro- and nano-grooves is expected to be responsible for high contact angle and high repellency in the above.

Comparing the microstructures in **Figures 12** and **13**, the amorphous carbon layer is physically laser-textured by controlling the ablation process within the beam spot to form the tailored surface structure even local. When laser-texturing the nitrogen supersaturated AISI420 die, a μm- to sub-μm sized island is formed on the terrace surfaces of microgrooves and the nanotextures are induced across these islands. This difference in induced microstructure by leaser-texturing comes from the ablation process of local heterogeneous microstructure and the chemical reaction after laser-irradiation, to be discussed later.

4. CNC-imprinting of textures to metals, polymers, and glasses

Both the DLC-coating die and the nitrogen supersaturated mold were utilized for CNC-imprinting the mother micro−/nano-textures on the dies into various work materials. The DLC-die was utilized in the cold and warm CNC-imprinting processes. The nitrogen supersaturated AISI420 mold was used in the hot CNC-imprinting process.

CNC-imprinting of textures to aluminum alloy. The CNC stamping system in **Figure 7** was utilized to imprint the star-shaped texture on the DLC-coating die onto the AA1060 aluminum alloy plate with the thickness of 1 mm and the as-rolled surface roughness. **Figure 14a** shows a star-shaped replica that is imprinted onto the aluminum plate by coining the micro-textured DLC punch. The original star-shaped emblem in **Figure 10** corresponds to the imprinted replica in **Figure 14a** in the mirror-inverse reflection. In parallel with the geometric imprinting from the microtextures

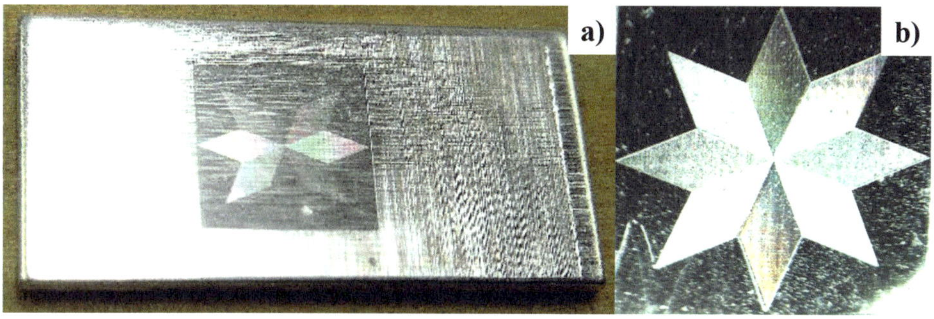

Figure 14.
Micro–/nano-textured AA1060 aluminum alloy plate by CNC-imprinting in cold. a) CNC-imprinting of the star-shaped emblem including the micro–/nano-textures onto the as rolled aluminum alloy plate with the thickness of 1 mm, and b) optical microscopy image of the star-shaped emblem.

on the DLC-die to the plate, the color-grating and surface plasmonic brilliance are also duplicated on this aluminum plate. This reveals that the intrinsic color grating and surface plasmonic brilliance to micro–/nano-textures on the DLC-die can be reproduced on any metallic work product surfaces by the means of accurate imprinting.

Surface decoration observed in **Figure 14** in a similar manner to **Figure 10**, reveals that the mother micro–/nano-textures are accurately imprinted onto the aluminum alloy plate and that the microstructure reproduced on the product surface is responsible for the similar surface decoration to the textured DLC-die. Let us investigate the microstructure of replica textures imprinted onto the plate. **Figure 15** depicts the SEM image on the micro–/nano-textures imprinted onto the AA1060 aluminum alloy plate with varying magnification from a) to d). A-zone in **Figure 15a** turns to be **Figure 15b**, B-zone in **Figure 15b** turns to be **Figure 15c**, and C-zone in **Figure 15c**

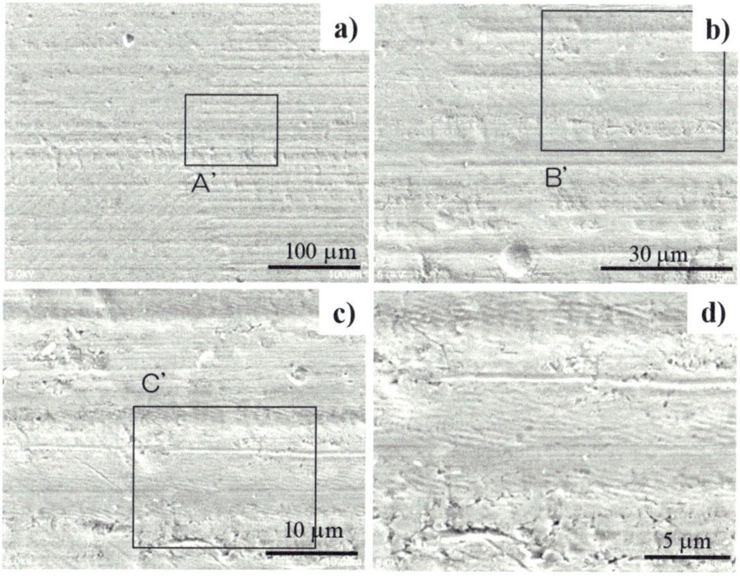

Figure 15.
SEM image on the micro–/nano-textures imprinted on the AA1060 aluminum alloy plate with varying the magnification from a) to d).

turns to be **Figure 15d**. In contrast to the mother micro–/nano-textures in **Figure 12**, the micro-textures in **Figure 15a** and **b** are just in the mirror-image reversal to the mother textures on the DLC-die in **Figure 12a** and **b**. Simply comparing **Figure 15c** and **d** to **Figure 12c** and **d**, the nanotextures are laterally aligned in **Figure 15c** and **d** against the longitudinal alignment of mother nanotextures in **Figure 12c** and **d**. This difference comes from the plastic flow of polycrystalline aluminum alloy with an average grain size of 20 μm into the DLC-die micro-cavities, to be discussed later.

CNC-imprinting of textures to PET. A PET (Poly-Ethylene Terephthalate) film was employed as a polymer work material for warm CNC-imprinting of mother textures on the DLC-die. In a similar manner to imprinting onto the aluminum alloy sheet, the DLC-die was indented into the PET film with a thickness of 0.2 mm. In this warm imprinting process, the thermocouple was embedded into the lower die. This die temperature was varied and optimized to search for the appropriate holding temperature (T_H) and holding time (T_i). In the following experiment, T_H = 290°, and T_i = 90 s.

Figure 16 depicts the center part of the star-shaped replica imprinted onto the PET film. Micro-edges of textures in the DLC-die were indented into PET film to form the micro-grooves. This proves that originally tailored textures on the DLC-die are accurately reproduced onto the polymer products by this warm imprinting process.

CNC-imprinting of textures to phosphorous glasses. A phosphorous glass type L-PHL2 has been widely utilized as a work for the fabrication of functional optical lenses and DOE. It is characterized by a low glass transition temperature of 381°C and low softening point of 440°C. The flow stress of this glass significantly decreases between these two temperature points; the hot mold stamping process makes use of this temperature range to shape the glass preform to an optical element and lens.

In this hot imprinting process, a cylindrical L-PHL2 preform with a diameter of 5 mm was used as a work and mold-stamped at 440°C for 90 s. **Figure 17** compares

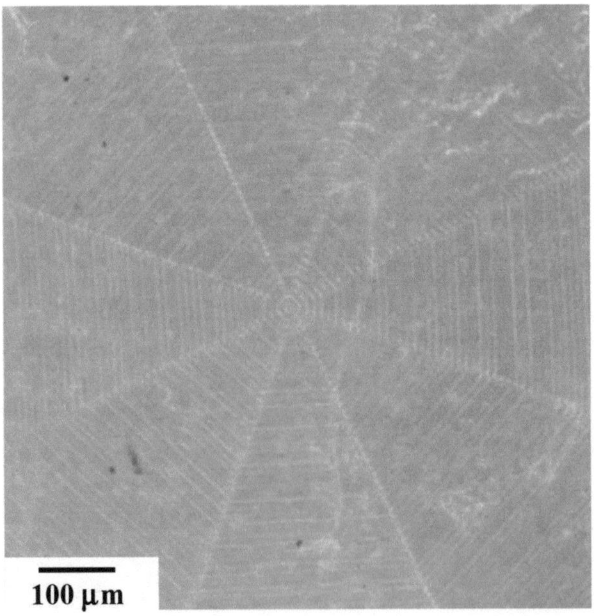

Figure 16.
Optical microscopy image on the micro–/nano-textures imprinted onto the PET film by compression of the DLC-die.

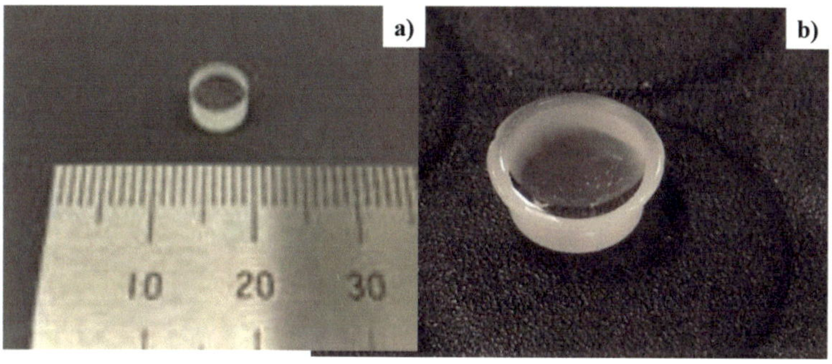

Figure 17.
A phosphorous glass type L-PHL2 preform before and after hot imprinting with the use of nitrided and laser-textured AISI420 mold.

the L-PHL2 preform before and after hot imprinting process. The micro–/nano-textures were shaped onto the preform surface in **Figure 17b**. Three-dimensional surface profilomer was also utilized to measure this surface profile.

Figure 18 shows the measured surface profile of L-PHL2 glass after CNC-imprinting. With comparison to the mother surface profile of DLC-die in **Figure 11**, both profiles are corresponding to each other with fairly good geometric compatibility in its peak-to-valley. This compatibility is dependent on the filling process of glass material work into the cavities of nitrided AISI420 mold. To be discussed in later, the imprinting process control has much influence on this filling process.

The wettability test was performed to investigate the surface property change before and after micro-texturing onto the L-PHL2 glasses by CNC-imprinting. As shown in **Figure 19a**, the swelling on the textured surface proves that the original hydrophilic glass surface changes to be hydrophobic. As measured in **Figure 19b**, the contact angle changed from 60° before imprinting to 114° after imprinting. The surface property of glasses can be controlled from the original hydrophilicity to the hydrophobicity by imprinting the tailored micro–/nano-textures.

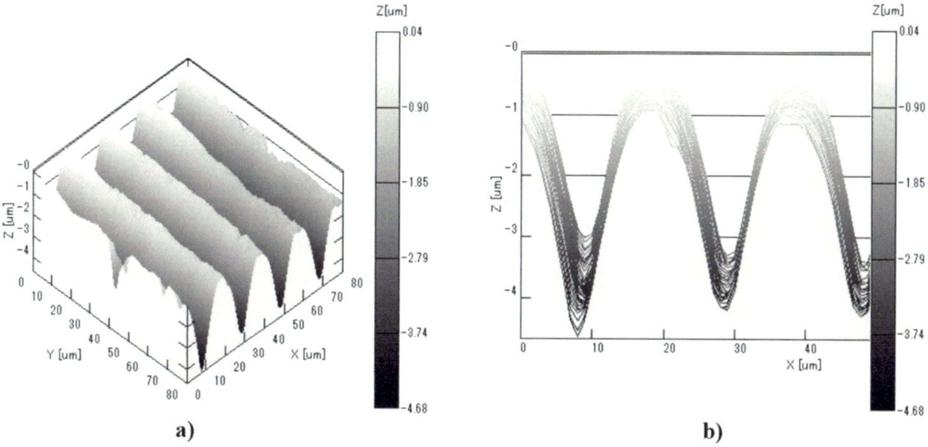

Figure 18.
The preform surface profile shaped by hot imprinting the laser-textured mold. a) Bird view of preform surface profile, and b) preform surface profile scanned in the x-axis.

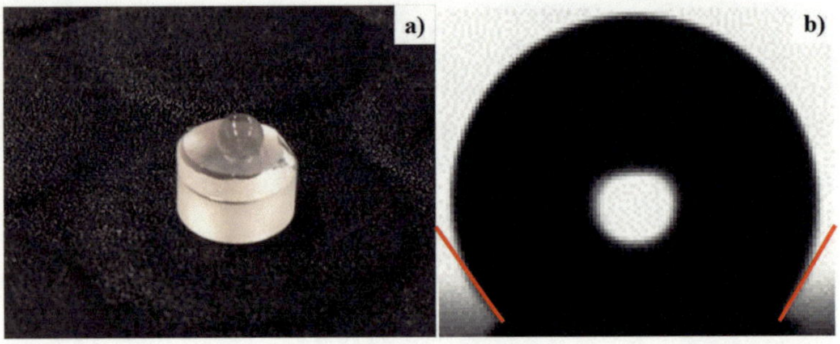

Figure 19.
Wettability on the textured L-PHL2 glasses by CNC-imprinting. a) A pure water droplet on the textured L-PHL2 glass surface, and b) hydrophobic surface with a contact angle of 114°.

5. Discussion

The femtosecond laser micro−/nano-texturing process to die substrate materials is driven by the ablation process, where each constituent matter of dies is vaporized by high-intensity beam spot during the laser irradiation duration of sub-ps without thermal effect. As studied in [4, 5, 27, 28], this high-intensity ablation process is described by the mass density effect on the micro−/nano-texturing. When texturing the CNT (Carbon Nano- Tube) coating, the removal depth in laser-drilling reached 10s–100 μm by a series of shots. In the case of laser-texturing of the glassy carbon, this removal rate under the same laser irradiation condition decreased down to 5–10 μm. On the other hand, the sp3-rich DLC and the diamond coatings needed more irradiation shots for drilling the depth of coatings [29]. This dependency on the mass density of carbon base coatings reveals that the power and fluence of laser irradiation must be tuned to each substrate materials.

In addition to this high intensity power deposition effect to ablation process, the polarization effect on the texturing is taken into account. After [30–32], the surface structuring in sub-mm range is significantly affected by the twisted laser beam through the polarization. The straight microgroove textures are formed and aligned under a selected polarization condition, while the nano-islands are formed and aligned under another condition. This polarization effect on the nanotexturing is also affected by the substrate material. As seen in the nano-texturing onto the DLC-die in **Figures 10** and **12**, the straight sub-mm sized nanogrooves are formed in regular alignment onto the DLC-die in almost all the polarization conditions. As already discussed in [7, 23] and seen in **Figure 12**, the nanogroove orientation is controllable by this polarization. On the other hand, the nano-island pattern is formed on the nitrogen supersaturated AISI420 die surface instead of formation of regularly aligned nanogrooves, as seen in **Figure 13**. To be noticed, the polarization condition is varied to form a mixture of nano-islands and nano-grooves as shown in **Figure 20**.

The straight nano-grooves are formed in the controlled orientation together with the nano-islands under the tuned polarization conditions. This nano-island formation by the polarization control suggests that the metallic nano-particle formation [33] and the carbon nano-dot deposition [34] are also tunable by locally twisting the laser beam.

In the cold imprinting process, the micro−/nano-textured cavity in the DLC-die is filled by the metallic work through the elasto-plastic flow. In the conventional metal forming, the metallic work is compressed by the loading sequence to fill into mm−/sub-mm sized cavities of dies. The filling volume fraction is determined when

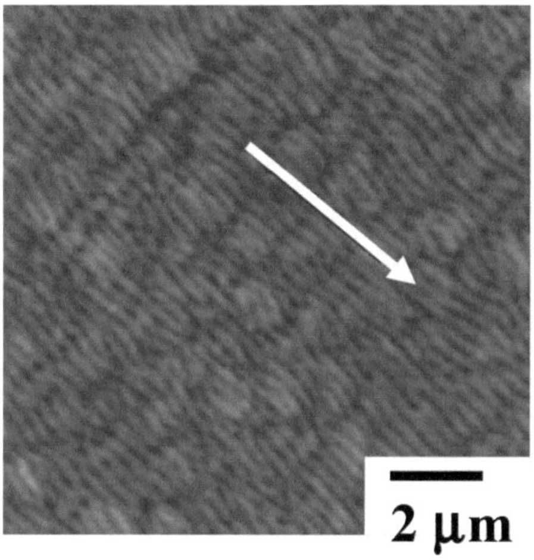

Figure 20.
Polarization effect on the formation of nano-islands mixed with nano-grooves under the controlled polarization when femtosecond laser texturing the nitrogen supersaturated AISI420 surface.

the applied stress is in equilibrium with the resistance flow stress of work materials. In the present micro-filling process into µm–/sub-µm sized grooves, the grain size of work has an influence on the polycrystalline plastic behavior. Let us describe the microscopic plastic flow of aluminum alloy work in filling the micro-textured and nano-textured grooves.

The filling process into two neighboring segments in the star-shaped emblem is considered in the following. Two segments on the DLC-die in **Figure 21** are imprinted onto the aluminum alloy plate in **Figure 21c**. As depicted in **Figure 21b**, the width of micro-groove terrace between two adjacent edges is 10 µm, smaller than the average grain size of 20 µm in the aluminum alloy work. Under the mechanical constraint by the grain boundaries, the work is thought to be plastically flown into the micro-terrace cavities of DLC-die by CNC-imprinting. Three-dimensional profilometer was utilized to describe this micro-filling process.

Figure 22 compares the surface profiles on the DLC-die and the textured aluminum alloy plate. The aluminum alloy work flew into the micro-cavity with the die terrace width (W_{die}) of 10.1 µm and the maximum terrace depth (H_{die}) of 0.7 µm as depicted in **Figure 22a**. **Figure 22b** measures the aluminum work after cold imprinting and releasing from the DLC-die. The convex micro-bump with the work width (W_{work}) of 9.2 µm and the work height (H_{work}) of 0.3 µm was formed after indenting the aluminum work into the concave terrace of DLC-die and releasing the work. Since $H_{work} < H_{die}$, this die terrace cavity was not fully filled by the work in this cold imprinting process. Since $W_{work} < W_{die}$, the imprinted work surface after releasing from the die, shrunk by 0.9 µm from the elasto-plastically deformed geometry during loading. As studied in [35], the spring-back of work occurs after releasing the work from the die by the fraction of elastic strains. The shrinkage of 9% in the above reveals that structural recovery in elasticity takes place with the material spring-back after releasing the work from the die. The micro-edges in the DLC-die indent into the work and play as a wedge to fix the work. The work is backward extruded into the die terrace cavity during this indentation of micro-edges into the work. This local plastic micro-flow of mono-grained aluminum work drives this micro-filling into the die terrace.

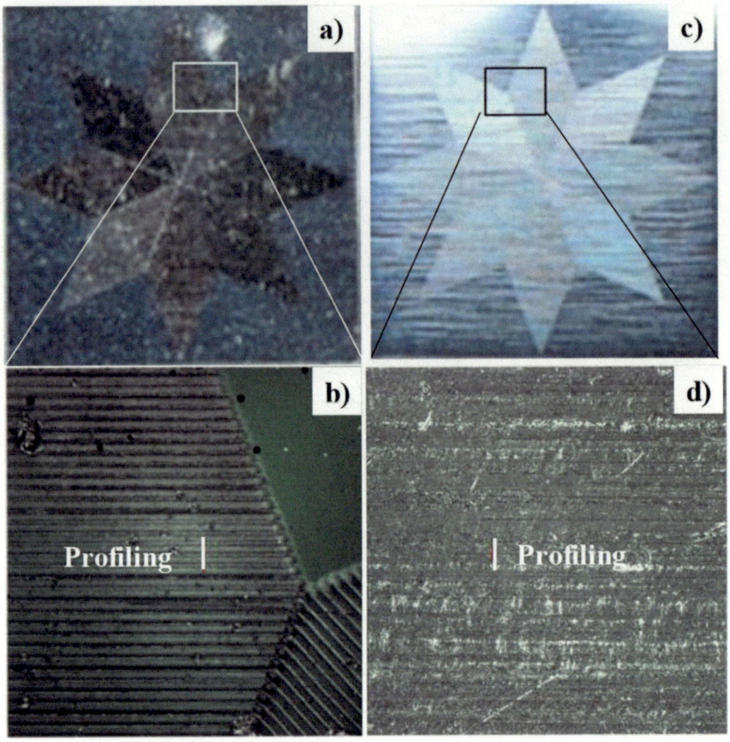

Figure 21.
Comparison of the micro-textured grooves on the DLC-die and the textured aluminum alloy plate.

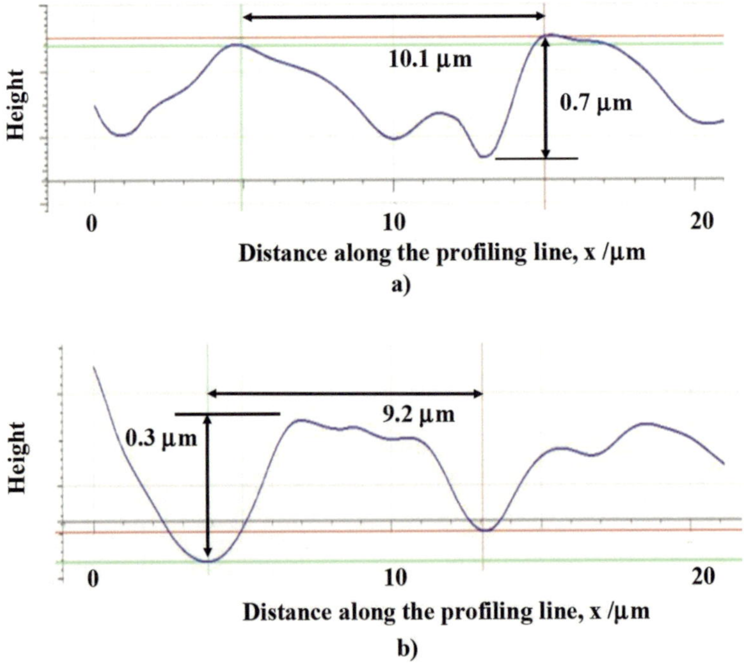

Figure 22.
Comparison of the surface profile between the micro-grooved terrace in the DLC-die and the shaped bump in the aluminum alloy plate by imprinting the DLC-die.

The side surfaces of extruded work peaks are partially in contact with the root of micro-edge and terrace surface in the DLC-die in **Figure 22**. The nanotextures on the DLC-die are thought to be imprinted into these side surfaces. As depicted in **Figure 15c** and **d**, the nano-textures were formed along the microgrooves.

In the hot CNC-imprinting of glass materials, they deform visco-elastically in the temperature range from the glass transition temperature to the softening one. Hence, the holding duration in stamping has much influence on the micro-filling of glasses into the DLC-die cavities.

The hot mold-stamping tests were performed to investigate the effect of holding duration on the filling process of glass work into concave textures. The inline measured time history of temperature is controlled as indicated in **Figure 23**. Owing to the IH heating, the heating transient to the specified holding temperature of 440°C had no over- and under-shooting steps; the measured temperature monotonously increased to T_H. The holding duration was directly controlled in this sequence. The cooling process was also controlled to be free from the thermal cracking on the contact surface between the die and the glass work.

The surface profiles of textured glass works were measured to calculate the average peak height (H) of glass works. **Figure 24** depicts the variation of H with increasing the holding duration (τ_H). H monotonously converged to the die cavity depth (H_{die})

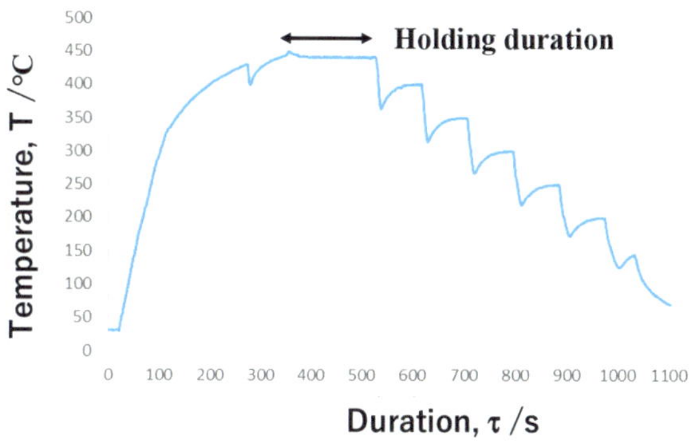

Figure 23.
A typically controlled temperature history for hot CNC-imprinting the textures onto the L-PHL2 preforms.

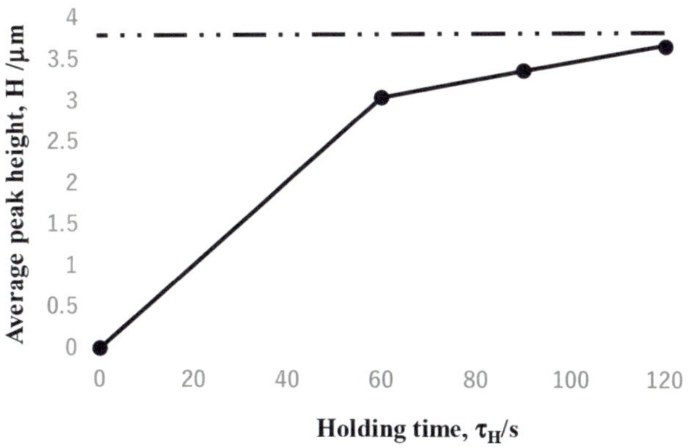

Figure 24.
Variation of the average peak height of textured L-PHL2 preform surface with increasing the holding duration.

with τ_H. This monotonic convergence of H to H_{die} reveals that the filling process of glass materials into the textured cavity in nitrided die is governed by the visco-elastic, time-dependent deformation of glass works.

When tailoring the fundamental periodic micro-texture in **Figure 11**, the monotonic filling process is sustained in the present hot imprinting procedure to fabricate the textured glass preform with the periodic textures in **Figure 18**. When the tailored textures are synthesized and formed from some periodic structures in **Figure 4b**, this filling process must be affected by the inhomogeneous deformation of glasses into the textured cavities in the nitrided die.

Let us investigate this topological effect of micro-textures on the nitrided die to the hot imprinting behavior. As depicted in **Figure 25a**, two periodic surface structures were synthesized and machined onto the nitrided die. Compared to the fundamental surface structure in **Figure 11**, the peak height and valley depth of laser-textured surface profile distribute on the die surface.

When hot mold-stamping the glass preform onto this die, a local filling process of glass material into each cavity in the die, advances in a different manner at each position. **Figure 25b** depicts the surface profile of imprinted glass preform. Four peaks of this surface profile (P_1, P_2, P_3, P_4) were formed by micro-filling of glass materials into four cavities (C_1, C_2, C_3, C_4) on the DLC-die in **Figure 25a**. If the filling process advances homogeneously in a similar manner to the imprinting process in **Figure 18**, the width and height of four peaks must corresponding to the width and depth of four cavities. As noticed in **Figure 25**, the geometric correspondence between micro-peaks in **Figure 25a** and micro-cavities in **Figure 25b** is not well-defined by the difference in micro-viscous flow of glass materials. Due to this inhomogeneous filling process in local, the maximum peak height of glass preform was reduced from 95% in **Figure 18** down to 81% in **Figure 25**. This reveals that the loading sequence during τ_H in **Figure 23** must be tailored to incrementally drive the local micro-viscous flow of glasses.

Finally, let us consider the application of femtosecond laser micro–/nano-texturing process with a direct imprinting process. As surveyed in [15], almost all

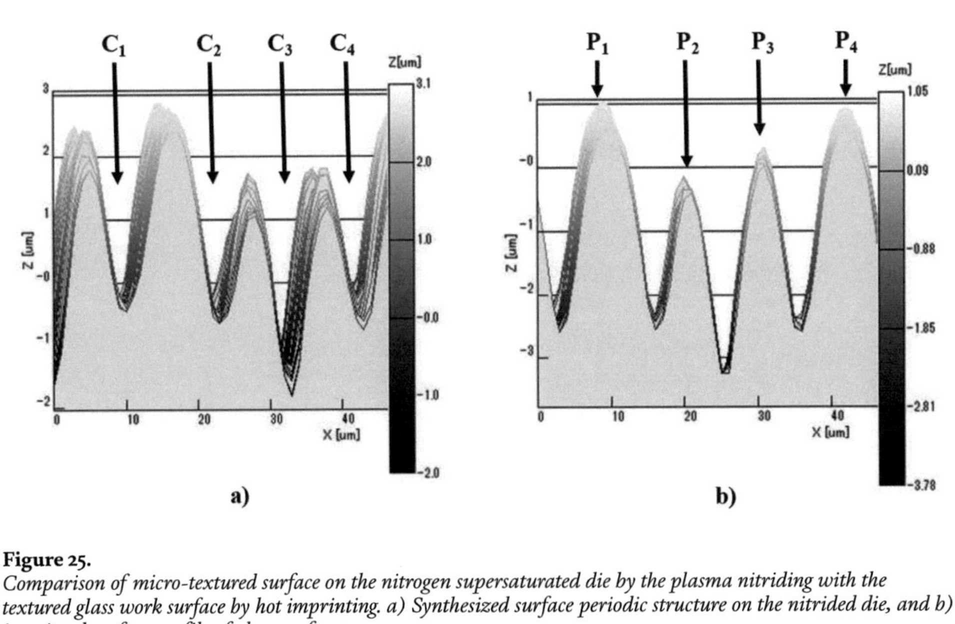

Figure 25.
Comparison of micro-textured surface on the nitrogen supersaturated die by the plasma nitriding with the textured glass work surface by hot imprinting. a) Synthesized surface periodic structure on the nitrided die, and b) imprinted surface profile of glass preform.

the tool surfaces can be DLC-coated with a significant thickness of more than 10 μm. Using this laser texturing technique, almost all the DLC coated tools are available as a mother die with the tailored micro–/nano-textures. Through the cold, warm, and hot imprinting processes, every metallic, polymer and ceramic product surfaces are decorated by the color-grating and surface plasmonic brilliance. In particular, the imprinting with the use of textured DLC-roll is effective to make large-area imprinting of textures onto metallic and polymer product surfaces. A surface decoration by the surface textures with a high aspect ratio is expected in cold imprinting instead of the polymer-based imprinting procedure [36].

The nitrogen supersaturated tool steel and stainless-steel dies are suitable for fine laser- and mechanical machining for micro–/nano-texturing. As stated in [37–39], PCD (PolyCrystalline Diamond) – chipped tools were available in fine texturing without the tearing of machined work-material surfaces and without the significant wear of PCD. This preciseness in dimension with robustness in texturing comes from the chemical stability of nitrogen supersaturated layer. Even in the femtosecond laser texturing, this chemical stability has an influence on the local ablation process by increasing the nitrogen solute content.

The hot imprinting of die textures is effective to change the original surface of glass preforms to be hydrophobic or super-hydrophobic during the mold-stamping of optical lens. In particular, a miniature lens in the endoscope and a micro-lens array in the detector are often covered by the surfactants such as the blood and body solution drops and the raindrops. This hydrophobicity works to prevent these lens surfaces from swelling with surfactants on them. In case of the meniscus lenses, its transparency is controllable by optimizing the microtexture depth.

6. Conclusion

Two-step laser-texturing base procedure is proposed to make surface decoration and surface property control on the metal, plastic, and glass products. The femtosecond laser micro–/nano-texturing works as the first step to form the tailored textures onto the hard-coating die and the surface-treated mold. Using the polygonal geometric models, various emblems, symbols, images, fonts, and pictures are textured onto the die surface. In particular, the textured DLC-die has sufficient hardness and chemical stability for long-term usage in cold and warm conditions. The nitrogen supersaturated tools by the low-temperature plasma nitriding plays as a reliable die to be micro–/nano-textured by using the femtosecond laser processing. Different from the homogeneous ablation process of DLC-die, the polarization in laser texturing must be controlled to form the nanotextures onto the micro-textured surfaces. This nitrided die has sufficient hardness and chemical stability in fine machining and in hot imprinting.

The CNC-imprinting is utilized as the second step to transcribe the mother textures on the die and mold them to the product surfaces. The cold imprinting of textures onto metallic product works well to reproduce the tailored micro–/nano-textured on the DLC-die. In this mechanical imprinting, the grain-size effect on the micro-plastic flow of metallic work has an influence on the accuracy of reproducibility in nanotexturing. The warm imprinting onto polymer product is also available to reproduce the micro–/nano-textures.

Hot imprinting procedure is needed to transcribe the laser-textured die surface into the glass preform. The hydrophobic die surface profile is imprinted onto the glass preform; the hydrophilic glass surface changes to be hydrophobic. This hot imprinting procedure is improved to increase the dimensional accuracy in transcription of micro–/nano-textures and to duplicate the tailored die surface profile with synthesized periodic structure.

The laser-textured DLC-die can be widely utilized to make large-area imprinting even onto the curved product surface and to form the textured product surface with a high aspect ratio. The nitrided stainless steel die is utilized to form the mother textured die surface by using the laser-processing and mechanical finishing processes. Various applications are acceptable to the present approach for micro–/nano-texturing of industrial and medical products.

Acknowledgements

The authors would like to express their gratitude to Okabe T. (LPS-Works Co., Ltd.), Yoshino T. (Komatsu-Seiki Kosakusho, Co., Ltd.), Hasegawa T., Miyagawa T. (Graduate School of Engineering, Shibaura Institute of Technology), and Kurozumi S. (Nano-Film Coat) for their help in experiments.

Conflict of interest

The authors declare no conflict of interest.

Author details

Tatsuhiko Aizawa[1*], Tadahiko Inohara[2], Yohei Suzuki[3] and Tomomi Shiratori[4]

1 Surface Engineering Design Laboratory, Shibaura Institute of Technology, Tokyo, Japan

2 LPS-Works, Co., Ltd., Tokyo, Japan

3 Komatsu-Seiki Kosakusho, Co., Ltd., Suwa, Japan

4 University of Toyama, Toyama, Japan

*Address all correspondence to: taizawa@sic.shibaura-it.ac.jp

References

[1] Ministry of Finance. Available from: https://www.asahi.com/articles/ASP916DKBP91ULFA00J.html [Accessed: February 9, 2021]

[2] Murugan P, Krishnamurthy M, Jaisankar SN, Samanta D, Mandal AB. Controlled decoration of the surface with macromolecules: Polymerization on a self-assembled monolayer. Chemical Society Reviews. 2015;**44**(10):3212-3243

[3] Pina-Estany J, Garcia Granada AA, Corull-Massana E. Injection moulding of plastic parts with laser textured surfaces with optical applications. Optical Materials. 2018;**79**:372-380

[4] Aizawa T, Inohara T. Multi-dimensional micro-patterning onto ceramics by pi-co-second laser machining. Research Report SIT. 2012;**56**(1):17-26

[5] Aizawa T., Inohara T., Pico- and femtosecond laser micromachining for surface texturing. Chapter 1 in Micromachining, IntechOpen, London, UK 2019: 1-24

[6] Michalek A, Qi S, Batal A, Penchev P, Dong H, See TL, et al. Sub-micron structuring/texturing of diamond-like carbon-coated replication masters with a femtosecond laser. Applied Physics A: Materials Science & Processing. 2020;**126**(144):1-12

[7] Aizawa T, Yoshino T, Suzuki Y, Komatsu T, Inohara T. Micro−/nano-texture surface decoration of metals via laser printing and precise imprinting. In: Proc. 13th AFGS (Dec., 2021; Hong Kong, China). 2021. pp. 79-86

[8] Aizawa T, Tamaki M, Fukuda T. Large area micro-texture imprinting onto metallic sheet via CNC stamping. Journal of Procedia Engineering. 2014;**81**:1427-1432

[9] Aizawa T, Yoshino T, Inohara T. Micro−/nano-texturing of aluminum by precise coining for functional surface decoration. Journal of Metals. 2020;**10**(1044):1-10

[10] Chaves J. Introduction to Non-imaging Optics. Second ed. Vol. 2015. CRC Press; 2015. pp. 213-240

[11] Aizawa T, Wasa K, Tamagaki H. A DLC-punch array to fabricate the micro-textured aluminum sheet for boiling heat transfer control. Journal of Micromachines. 2018;**9**(147):1-10

[12] Aizawa T, Inohara T, Wasa K. Fabrication of super-hydrophobic stainless steel nozzles by femtosecond laser micro−/nano-texturing. International Journal of Automation Engineering. 2020;**14**(2):159-165

[13] Hosford WF. Metal Forming: Mechanics and Metallurgy. 4th ed. UK: Cambridge University Press; 2014

[14] Sanditov DS, Ojovan MI, Darmaev MV. Glass transition criterion and plastic deformation of glass. Physica B: Condensed Matter. 2020;**582**:411914

[15] Aizawa T. Advanced technologies for die materials and manufacturing. Bulletin, JSTP. 2021;**4**(41):291-296

[16] Aizawa T., Low temperature plasma nitriding of austenitic stainless steels. Chapter 3 in Stainless steels and alloys. IntechOpen, London, UK 2019: 31-50

[17] Aizawa T, Yoshino T, Suzuki Y, Shiratori T. Anti-galling cold dry forging

of pure titanium by plasma carburized AISI420J2 dies. Journal of Applied Sciences. 2021;**11**(595):1-12

[18] Aizawa T, Yoshino T, Suzuki Y, Shiratori T. Free-forging of pure titanium with high reduction of thickness by plasma carburized SKD11 dies. Journal of Materials. 2021;**14**(2536):1-12

[19] van Driel HM, Sipe JE, Young JF. Laser-induced periodic surface structure on solids: A universal phenomenon. Physical Review Letters. 1982;**49**:1955-1958

[20] Kawase T. Superhydrophobic surface. SEN'I Gakkai. 2009;**65**:200-207

[21] Gečys P, Vinčiūnas A, Gedvilas M, Kasparaitis A, Lazdinas R, Račiukaitis G. Ripple formation by femtosecond laser pulses for enhanced absorptance of stainless steel. Journal of Laser Micro/Nanoengineering. 2015;**10**:129-133

[22] Hasegawa T, Aizawa T, Inohara T, Yoshihara S-I. Mold stamping of optical glasses using super-hydrophobic surface die by micro/nanotexturing. Journal of JSTP. 2019;**60**(698):75-79

[23] Aizawa T, Inohara T. Imprinting of Micro−/Nano-Textures into Metals and Alloys with Use of the Laser-Printed DLC-die. Proc 25th ESAFORM (Braga, Portugal; April 27th, 2022). Vol. 2022. p. 96

[24] Deng X, Kousaka H, Tokoroyama T, Umehara N. Thermal stability and high temperature tribological properties of a-C:H and Si-DLC deposited by microwave sheath voltage combination plasma. Tribology Online, JST. 2013;**8**(4):257-264

[25] Aizawa T, Inohara T, Wasa K. Femtosecond laser micro−/nano-texturing of stainless steels for surface property control. Journal of Micromachines. 2019;**10**(512):1-10

[26] Onda T. Physics of wettability and water repellency of fractal-structure surfaces. Journal of the Vacuum Society of Japan. 2015;**58**(11):424-430

[27] Aizawa T, Inohara T. Micro-texturing onto glassy carbon substrates by multi-axially controlled pico-second laser machining. In: Proc. 7th ICOMM (2012, March 12; Evanston, USA). 2012. pp. 66-73

[28] Aizawa T, Inohara T. Multidimensional micro-patterning onto ceramics by pico-second laser machining. Research Report SIT. 2012;**56**(1):17-26

[29] Aizawa T., Inohara T., Yoshino T., Suzuki Y., Shiratori T., Laser treatment of CVD diamond coated punch for ultra-fine piercing of metallic sheets. Ch. 4 in Engineering Applications of Diamond, Intech Open, UK, London 2021:71-94

[30] Maalouf M., Khalil A. A., Maio Y. D., Papa S., Sedao X., Dalix E., Peyroche S., Guignandn A., Dumas V., Polarization of femtosecond laser for titanium alloy nano-patterning influences osteoblastic differentiation. Journal of Nanomaterials 2022; 12 (10), 1619: 1-12

[31] Bai F, Li H-J, Huang Y-Y, Fan W-Z, Pan H-H, Wang Z, et al. Polarization effects in femtosecond laser induced amorphization of monocrystalline silicon. Chemical Physics Letters. 2016;**662**(1):102-106

[32] Zhu J-T, Shen Y-F, Chen X, Yin G, Chen D-Y, Zhao L. Effect of polarization on femtosecond laser pulses structuring silicon surface. Applied Surface Science. 2006;**252**(8):2752-2756

[33] Almeida JMP, Ferreira PHD, Manzani D, Napoli M, Ribeiro SJL. Metallic nanoparticles grown in the core of femtosecond laser micromachined wave guilds. Journal of Applied Physics. 2014;**115**(19):221-226

[34] Abere MJ, Chen C, Rittman DR, Kang M, Goldman RS, Phillips JD, et al. Nanodot formation induced by femtosecond laser irradiation. Applied Physics Letters. 2014;**105**:163103

[35] Reddy BL, Rao BC, Reddy PR, Reddy PVRR. A review on spring back in metal forming. IHERT. 2014;**3**(1):21-25

[36] ENOS. Available from: https://www.eneos.co.jp/english/company/rd/intro/function/print.html [Accessed: May 14, 2022]

[37] Aizawa T, Fukuda T. Microstructure and micro-machinability of plasma nitrided AISI420 martensitic stainless steels at 673 K. In: Top-5 Cont. to Mater. Sci. 6th ed. Avid Science; 2019. pp. 2-23

[38] Aizawa T, Morita H, Fukuda T. High machinability of plasma-nitrided HPM80 dies at 673K by PCD-tools for hot mold-stamping. Procedia Manufacturing. 2020;**47**:725-731

[39] Aizawa T, Morita H, Fukuda T. Nitrogen Supersaturation into AISI420 Mold for Precise Machining. Proc. 25th ESAFORM (Braga, Portugal; April 28th, 2022)2022. p. 96

Chapter 5

Ablation of Materials Using Femtosecond Lasers and Electron Beams

Mehra S. Sidhu and Nitish Dhingra

Abstract

The advancements in producing interactions of concentrated energy fluxes, such as femtosecond lasers and high-energy electron beams with the absorbing substances, have facilitated new discoveries and excitement in various scientific and technological areas. Since their invention, significant improvements in temporal, spatial, energetic, and spectroscopic characteristics have been realized. Due to the ultrashort pulse width and higher intensity (10^{12} W/cm^2), it is possible to ablate the materials with negligible damage outside the focal volume, thereby allowing the treatment of biological samples, such as live cells, membranes, and removal of thin films, as well as bulk materials for many applications in diverse fields, including micro-optics, electronics, and even biology under extremely high precision. Since most biological systems are transparent toward the NIR spectral range, it follows the nonlinear multi-photon absorption interaction mechanism. In contrast, the electron beam follows linear absorption mechanism for material modifications even at lower energies. For realizing the fs-laser nano-processing in material applications, such as silicon microchips, or in biology like retinal cells, it is crucial to find a way to deliver these pulses precisely at the site of action and enhance the selectivity. The utilization of electron beams in material modification has also been exercised widely to attain nanoscale precision. In the next section, biological materials, such as cornea, retina, and silk, are discussed.

Keywords: ablation, biosensors, electron beams, femtosecond lasers,, nano-processing

1. Introduction

Ablation could be defined as the removal of material from the surface of an object by concentrated energy fluxes, such as femtosecond lasers and high-energy electron beams, by vaporization, or other erosive processes. It is a complex phenomenon that occurs upon reaching a specific energy threshold that causes measurable damage and material removal from a surface.

Surface modification of materials and devices is important for many applications, such as enhancing biocompatibility [1], promoting adhesion [2], improving wear resistance [3], preventing corrosion [4], providing hydrophilicity, hydrophobicity and electrical insulation or conducting properties, and generating antimicrobial and

IntechOpen

antibacterial surfaces [5, 6]. Considerable advancements have been made in developing new surface modification techniques. This chapter is focused on the laser ablation (LA) and electron beam ablation (EBA) techniques, which are quite popular for material ablation. The first part of this chapter discusses the details of the LA technique, whereas the latter part is focused on the EBA technique.

The ablation threshold, F_{th} (J/cm^2), indicates the minimum radiant exposure required to achieve effective ablative material removal. Meanwhile, the threshold fluence determines the possible precision of laser effects used for ablation and the etching of materials [7]. A low ablation threshold is generally favorable for irradiating a material with an ultrafast laser pulse to minimize the possible photo-induced damage close to the ablation area [8]. Irradiation by an intense ultrafast laser beam results in multi-photon excitation of a target material. Perhaps, the absorbed energy is transported to the electrons without thermal diffusion to the surrounding material because of the shorter pulse width of the incident pulse in comparison to the vibrational relaxation time constant of several picoseconds. As a result, thermal damage to the adjacent tissues could be minimized, and the biological tissue remains unaffected by the subsequent photo-induced mechanical shock process. This makes fs laser-assisted surgical processes non-thermal. In the vicinity of focus, the formation of a high density of free electrons could result in a local plasma formation in the targeted materials. This hot plasma formation results in permanent removal of material, even inside a cell within a sub-micron size [9].

Before discussing the details of laser ablation, it is imperative to discuss pulsed-laser interactions with biomaterials. The biomaterials discussed here, are the materials having low thermal/electrical conductivity and no free electrons. Common materials, such as fused silica, sapphire, bone, retina, silk, cornea, heart-tissue, and fall under this category [10]. In the next section, biological materials, such as cornea, retina, and silk, are discussed.

2. Laser interactions with biomaterials

The interaction of light with matter can occur through several different nonlinear processes, such as two-photon or multi-photon absorption, second harmonic generation, or Raman scattering. It is useful in deep-tissue two-photon imaging. Moreover, if irradiance is high enough, any material can be machined to very high precision by using ultrashort laser pulses. This interaction is independent of linear absorption properties of the materials and which would otherwise be transparent to the laser wavelength. Ablations to micron scale precision with minimal collateral damage to the rest of the material are achieved by faster removal of material than the rate of heat conduction to the bulk.

In present clinical practices, pulses longer than a few tens of picoseconds have been utilized. It can cause damage to the dielectrics involving heating of seed electrons and transfer of this energy to the lattice. This damage occurs via conventional thermal deposition resulting in the melting and boiling of dielectric material (**Figure 1(a)** and **(b)**). Because the energy is transferred through thermal conduction, this model predicts a square root ($\tau^{1/2}$) dependence of the threshold fluence (F_{th}; energy/area) upon pulse duration (τ) (**Figure 2**) [13].

With these conventional lasers, the material is removed by thermal ablation, wherein the material is locally heated to near the boiling point. Since the boiling point of dielectrics is very high ~typically 1000°C, this ablation is coupled with a strong

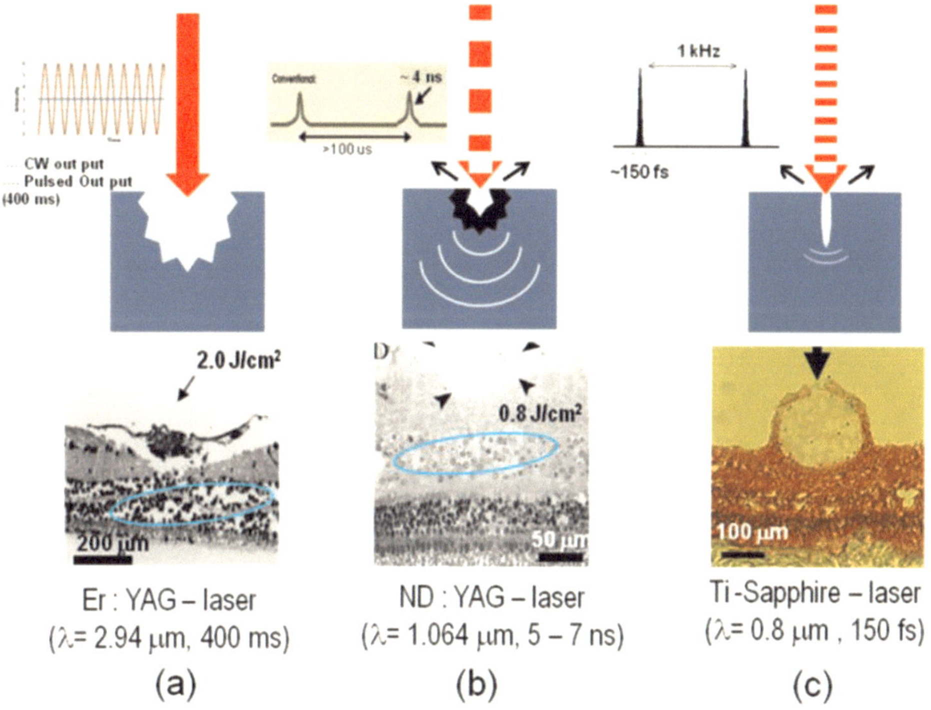

Figure 1.
(a) Ablations of retinal-tissue with 2.94 μm millisecond pulses (2 J/cm², 400 ms), (b) retinal ablations with 1.064 μm nanosecond pulses (0.8 J/cm², 5–7 ns), and (c) retinal blood vessel ablation with 0.8 μm femtosecond pulses (150 fs) [11, 12].

thermal shock transferred to the adjacent material. This thermal shock often results in cracking of the adjacent material in an uncontrolled manner. These effects can be observed in **Figure 1(a)** and **(b)**, where the ablations of porcine retina using a conventional Er: YAG (λ = 2.94 μm) and ND: YAG laser (λ = 1.064 μm) are shown [11, 12]. Meanwhile, fs-pulses (800 nm) having sufficient intensity for multi-photon ionization throughout its beam waist ablate the retinal vessel walls in an extremely controlled manner (**Figure 1(c)**). The heat transfer into the surrounding material was minimal and also no thermal shock-induced cracking was observed. It could be inferred that the laser damage threshold for dielectrics with lasers having pulse durations less than a few picoseconds to femtosecond does not follow the proportionality rule of square dependence [14–16]. In the next section, various ionization mechanisms operating in biological materials are briefly discussed.

3. Mechanism of ionizations

The changes in surface morphology and deviation from the $\tau^{1/2}$ dependence of the damage threshold on pulse duration are predicted by a rapid ionization mechanism. As shown in **Figure 4**, field-induced multi-photon ionization produces free electrons that are then rapidly accelerated by the laser pulse. This could be further described as avalanche ionization and tunnel ionization. As shown in **Figure 3**, by absorbing the incident photons, the kinetic energy of a free electron becomes sufficiently high and a

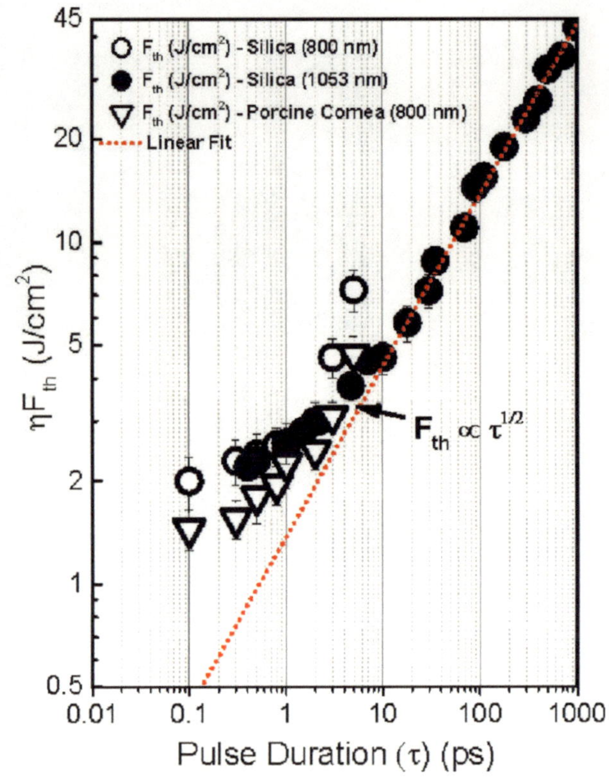

Figure 2.
Threshold dependence on incident energy density on laser pulse duration. A square root dependence is postulated for pulse durations between picoseconds and microseconds. The parameter is tissue-dependent and is expressed in J/cm² [13].

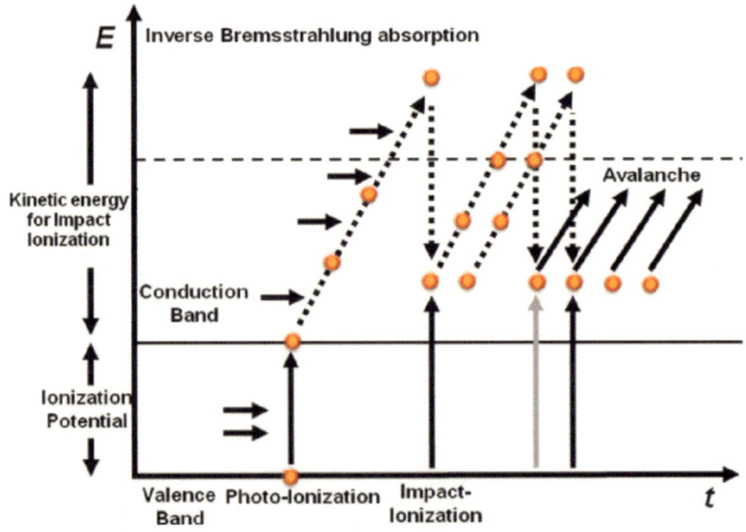

Figure 3.
Interplay of photo-ionization, inverse bremsstrahlung absorption, and impact ionization in the process of plasma formation. Recurring sequences of inverse bremsstrahlung absorption events and impact ionization lead to an avalanche growth in the number of free electrons [19].

part of the energy may be transferred to the bound electrons by collisions to overcome the ionization potential, and hence, produce two free electrons. This is referred to as "*collisional impact ionization*" [16, 19–22]. Subsequently, the free electrons absorb photons and produce more free electrons from the bound electrons. Such a series of the impact ionization process is called "*avalanche ionization*" [16, 17]. The avalanche ionization strongly depends on free electron density and is assumed to vary linearly with the laser intensity [23]. Its efficiency is determined by competition between energy gain through *inverse bremsstrahlung absorption* and loss in energy through the phonon emission. Avalanche ionization is responsible for the ablation of wide bandgap materials at the laser intensities below 10^{12} W/cm^2.

When mode-locked fs-laser pulses with intensity higher than 10^{13} W/cm^2 interacts with dielectrics, initiation of multi-photon ionization (MI) takes place. As shown in **Figure 4**, several (n) photons with the energy $h\nu$ having wavelength (λ) coherently strike the bound electron, acting like a single photon of $nh\nu$ (energy) at the wavelength of λ/n. It results in significantly higher photon flux (>10^{31} cm^{-2} s^{-1}) that allows a valance band electron to be freed. Meanwhile, absorbing several photons until the total energy would exceed the values of ionization potential [24, 25]. This absorption process is achieved through the meta-stable quantum state(s).

In Q-switched nanosecond laser pulses associated with low intensities, the initial process for the generation of free electrons is supposed to be thermionic emission, that is, the release of electrons due to thermal ionization. Rather, energies and temperature are usually higher in the case of Q-switched laser pulses because of the associated increment in threshold energy of ablations. Thus, laser-induced distortions with ns-pulses are often accompanied by nonionizing side effects [18].

At intensities above 10^{12} W/cm^2, multi-photon absorption becomes considerably strong and even the seed electrons are not required to initialize the ionization process [10, 26]. However, tunnel ionization should be considered when the intensities are higher than 10^{15} W/cm^2. The tunneling ionization is the process in which the strong

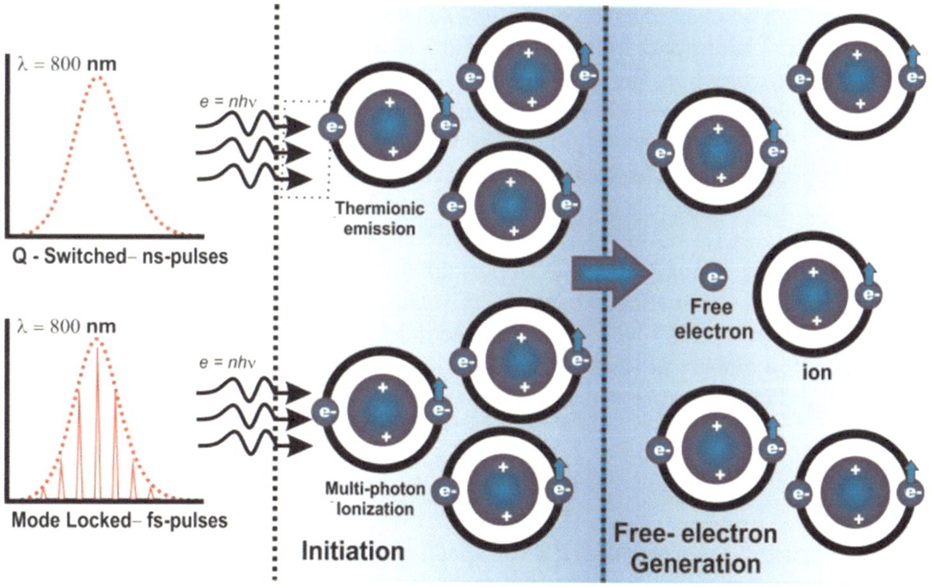

Figure 4.
Schematic representation of multi-photon ionization [17, 18].

incident field suppresses the coulomb potential to allow the tunneling of a bound electron to a free state. The coulomb potential that describes the interaction between two-point charges acting along the line connecting the two charges can be expressed by the following expression; $V_{coulomb} = {q_1 q_2}/{4\pi\varepsilon_o r}$, where r is the distance between two ions, q_1 and q_2 represent electric charge in coulombs carried by 1 and 2, respectively, and ε_o is the electrical permittivity of the space. In solids, for example, the bound electron in the valence band is excited through either multi-photon or tunneling to the conduction band and becomes quasi-free.

Keldysh developed the theory that describes the ionization of electrons in condensed matter by intense laser fields [27]. Keldysh parameter used to predict the mechanism that plays a significant role in the ionization process is defined as:

$$\gamma = \sqrt{\frac{m\omega_o^2 c n_0 \varepsilon_o E_{gap}}{e^2 I}} \tag{1}$$

where ω_o denotes the incident laser light frequency, m is the reduced mass of the electron, and E_{gap} is the band gap of the material. γ can be qualitatively viewed as the ratio of the incident laser frequency and field strength *(I)* [28]. For relatively weak fields with high frequencies (large γ), multi-photon ionization is more important since electrons have less time to tunnel through the only moderately suppressed coulomb potential than in the small γ case (strong field with low frequency).

3.1 Laser-induced breakdown and plasma formation

When obtaining power densities of the incident laser field (>10^{12} W/cm^2) is equivalent to or more than the coulomb field that binds the electron to its ionic core, the atomic coulomb force of similar magnitude was exerted on to the valance band electron and can be excited to a free state. The recurring sequence of *inverse bremsstrahlung* events and impact ionization leads to an avalanche growth in the number of free electrons. Meanwhile, the irradiance should be high enough to compensate for the losses of free electrons through diffusion out from the focus and through recombination [15, 16, 19, 20]. The energy gain at the vicinity of focus was more rapid than the energy loss by collisions with surrounding particles in media occurring without simultaneous absorption of a photon. These "cascade ionizations" in dielectrics followed by plasma formation leads to a phenomenon called *laser-induced optical breakdown* (LIOB), which can result in permanent material modification. Net free electron density is assumed to saturate at the critical density, at which modification/ablation of material takes place. For femtosecond lasers, critical density is selected as the free electron density at which plasma oscillation frequency equals to the laser frequency, and can be expressed as:

$$n_{cr} = \frac{\pi m_e c^2}{e^2 \lambda^2} \tag{2}$$

where m_e is the electron mass, c is the speed of light, e is the electron charge, and λ is the laser wavelength. n_{cr} is about 10^{21} cm^{-3} for 800 nm wavelength. At critical electron density, transparent dielectric materials become opaque.

4. Estimation of multi-photon absorption coefficients for biomaterials

Theoretically, the multi-photon absorption is determined by the following equation,

$$\frac{dI}{dz} = \alpha_1 I + \alpha_2 I^2 + \alpha_3 I^3 + \alpha_4 I^4 + \alpha_5 I^5 \qquad (3)$$

where I is the intensity of incident light beam propagating along the z-axis. The coefficients α_1, α_2, α_3, α_4, α_5 are one-, two-, three-, four- and five-photon absorption coefficients for a given medium, respectively. For a biomaterial, such as spider silk fiber of diameter D_0, we define the term absorption, A = (dI/I) $(1/D_0)$, which, using Eq. (3) becomes a polynomial of fourth-order, $A = \alpha_1 + \alpha_2 I + \alpha_3 I^2 + \alpha_4 I^3 + \alpha_5 I^4$. This mixed fit was used to fit the experimental data. A comparison of mixed fit with cases of pure 2-, 3-, 4-, and 5-photon absorption processes are also shown for comparison. Our fit analysis suggests that as the intensity of the pulses increases, the absorption is dominated by progressively higher-order multi-photon processes. The existence of mixed multi-photon processes for some intensity is also likely [20].

Another technique, "Z-Scan" can also be used to determine multi-photon absorption. The single silk fiber was illuminated with sub-10 fs laser beam at 85 MHz rep rate. The Rayleigh range was approximately, Z_R = 25 μm and the silk sample was typically D_0 = 1–4 μm in diameter. Thus, we satisfy the thin-sample ($Z_R >> D_0$) condition for the Z-scan measurements. The total transmission was measured by collecting the maximum light and focusing it on a photodiode, while the sample was scanned at low power and high power. The sample was translated along the beam path using a computer-controlled stage that allowed us to collect the data at about 2 μm spacing. The normalized data for high power was obtained by subtracting the low power data to minimize any variation due to diffraction effects from silk fibers as shown in **Figure 5**. We fitted our data with the nonlinear transmission function described in Ref. [20], and reference therein. The fit function for 2-photon was, $2(x) = Log(1 + C_2/(1 + x^2))/(1 + x^2)$, with a fitting parameter C_2. The fit function for the 3-photon absorption was, $T3(x) = Sinh^{-1}\left[\frac{C_3}{1+x^2}\right]/[1 + x^2]$. The fit function for 4-photon absorption was, $\mathrm{T4[x]} = \mathrm{H2F1}\left[1/3,1/3,4/3, -\left(\frac{1*C_4}{1+x^2}\right)^3\right]$, with C_4 was a fit parameter. The hypergeometric function H2F1 was computed in Mathematica. Our data fitted better with 4-photon absorption fits. This was consistent with our other measurements of nonlinear transmission [20].

5. Laser interaction mechanisms

On the implication of laser light into the transparent media, such as biological tissues, as well as viscoelastic like spider silk, various interaction mechanisms were facilitated. Specifically, the tissue characteristics along with the laser parameters are contributing to this diversity. Most important amongst them are optical properties of tissues that include the coefficient of reflection, absorption, and scattering. Altogether, these parameters determine the total transmission of the tissues and viscoelastic at a certain wavelength.

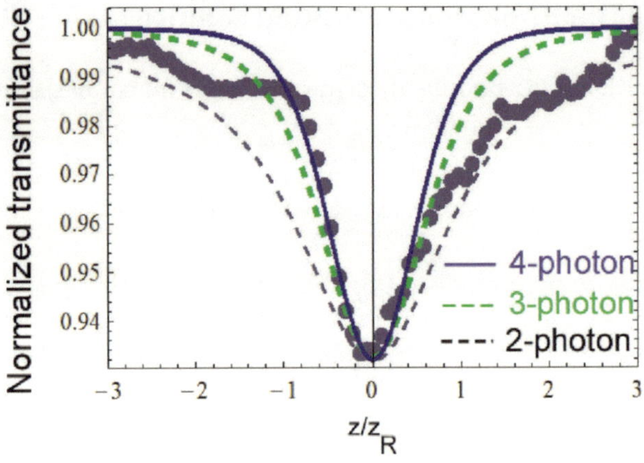

Figure 5.
Z-scan data from a single fiber (~150 GW/cm²) along with equations with 2, 3, 4-photon. The mismatch near edges was attributed to diffraction losses from the edges of the silk filament. Data taken from Sidhu et al. [20].

In contrast, the following parameters that affect the laser ablation, are given by the laser radiations themselves: applied energy, wavelength, exposure time, focal spot size, and energy density. Amongst them, exposure time is a very crucial parameter when selecting a certain type of interaction.

Laser-tissue interactions can be subdivided into five different regimes depending on the laser power and the exposure time. **Figure 6** gives us a rough delineation of the interaction regimes. Two lines represented diagonally show the fluences (energy per unit area) at 1 J/cm^2 and 1 kJ/cm^2, respectively.

According to this plot, the time scale can be roughly divided into five parts: continuous wave or exposure times >1 s indicates photochemical interactions, 1 min

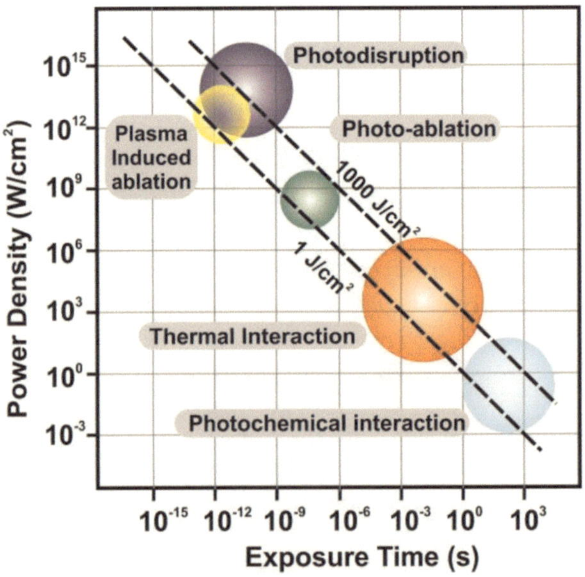

Figure 6.
A double logarithmic map of five basic laser-tissue interactions. The circle gives only a rough estimate of the associated laser parameters [18].

down to 1μs indicates thermal interactions, 1 μs down to 1 ns indicates photo-ablation, and < 1 ns indicates plasma-induced ablation, as well as photodisruption. The difference between the latter two is ascribed to different energy densities [18]. Precise laser surgery can be achieved when the desired target tissue, such as retina or ocular-tissue, is exposed to the optimal laser conditions following the subsequent interaction pathway.

5.1 Optical absorption and light propagation

Optical absorption takes place when the frequency of incident laser irradiation matches with an electronic excitation frequency of a molecule [18]. The typical absorption coefficient estimated by inverse Monte Carlo analysis for bovine cornea and retina at 800 nm was $\sim 0.001\ cm^{-1}$ and $2.5\ cm^{-1}$, respectively [29].

The coefficients for 2, 3, 4, and 5 photon absorptions for spider silk were $\alpha_2 = 1 \times 10^{-2}$ cm/GW, $\alpha_3 = 2 \times 10^{-5}\ cm^3/GW^2$, $\alpha_4 = 4 \times 10^{-6}\ cm_5/GW^3$, and $\alpha_5 = 5 \times 10^{-7}\ cm^7/GW^4$, respectively (**Figure** 7) [20, 30]. Previously, a 3-photon absorption process was observed in the silk-fibroin solution [31], and enhanced 3- and 4-photon absorption was reported for amyloid fibril solution [30].

In the case of biological tissues, such as retina and cornea-tissue, neither water nor macromolecules are present, it absorbs the light strongly in the near-infrared range (roughly between 600 to 1200 nm), this spectral range is considered a "therapeutic window." Laser radiation in this window can penetrate deeper into biological tissues with lower loss during treatment.

Light propagation inside the retina or corneal tissues can be attributed to their absorption and scattering properties. The effective attenuation coefficient (A_{eff}) quantifies how deep the laser light can penetrate, and is expressed as:

$$A_{eff} = \frac{1}{\sqrt{3\alpha_a(\alpha_s + (1-g))}} \tag{4}$$

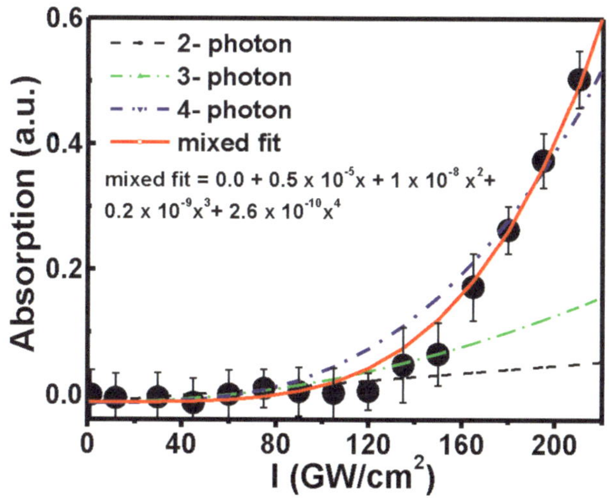

Figure 7.
Nonlinear absorption coefficient of silk fiber versus input peak intensity. Theoretical fits for pure 2-, 3-,4- photon absorption are compared with a mixed fit up to 4th order polynomial [20].

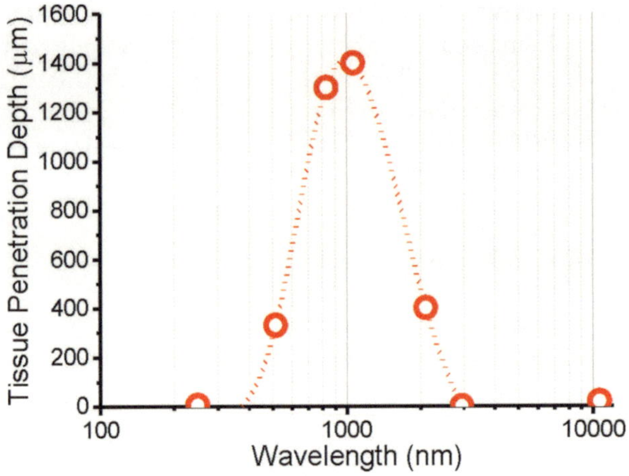

Figure 8.
Laser penetration depth in tissue at different wavelengths [21, 27].

where α_a and α_s are absorption and scattering constants. g is the coefficient of anisotropy which is a tissue-dependent parameter. With g = 1 denotes perfectly forward scattering, such as cornea, the effective penetration depth (L_{eff}) of incident light is.

$$L_{eff} = \frac{1}{\alpha_{eff}} \tag{5}$$

Figure 8 Illustrates the approximate penetration depth in tissues of different wavelengths, while both,α_a and α_s are taken into account.

The selection of appropriate wavelength and pulse energy followed by linear or nonlinear propagation of laser irradiations into the matched tissue target allows unprecedented precision to achieve surgical effects.

5.2 Photochemical interactions of fs-lasers

It is a well-known fact that light can induce chemical effects or reactions within biological tissues. In ophthalmology, photochemical interactions can play a vital role in dye-assisted photodynamic therapies (PDT). During PDT, photosensitizers were injected into the target tissues. Laser irradiations of specific wavelength trigger the photochemical reaction at low power densities (~1 W/cm^2) and long exposure time in order of seconds to continuous wave (CW) [18, 32]. It results in toxic reaction products that cause irreversible destruction of target cell structures. The reaction could be the generation of reactive oxygen species from the interaction of light, oxygen, and photosensitizers, such as verteporfin, and indocyanine green (ICG). These were commonly used to treat choroidal or corneal neovascularization, that is, abnormal blood vessels underlying retina cell layers or sprouting in the corneal stroma, respectively. However, severe adverse reactions are associated with these photosensitizers, including visual disturbance, stromal haze, and lipid degeneration injection site reactions even after months of treatment [33]. Several joules of energy (~150 J/cm^2) continuously deployed into corneal stroma allows long-term reduction of neovascular structure, but can also cause corneal scarring or thermal injuries [34, 35].

We suppose that without the use of chemical agents, the intensity of femtosecond laser pulses could be used to manipulate abnormal blood vessels lower than these conventional photodynamic treatment procedures. We will discuss the selective removal of corneal and retinal blood vessels in the next section of this chapter. The precise cutting and manipulation of ultrafine fibers are also discussed.

5.3 Thermal interactions

Thermal effects can be induced by both CW and pulsed-laser irradiations. The rise in local temperature could be attributed to the absorption of photon energy by protein, pigment, or bound water molecules. On the basis of pulse duration and peak values, different thermal effects can be achieved in tissues, including coagulation, vaporization, carbonization, and melting. Moreover, with the rise in temperature to 50°C, the enzyme activity in tissue cells is reduced, energy transfer slows down, and cell repair mechanisms are disabled. At about 60°C, protein denaturation occurs, leading to more immediate cell necrosis and tissue coagulation. Therefore, photocoagulation is the general procedure to treat retinal disease, such as proliferative diabetic retinopathy. Meanwhile, traditional lasers employed for photocoagulation (spot size: 0.1–3 mm, pulse duration: millisecond) creates a heat wave that spreads beyond the focal volume causing inevitable collateral damage to underlying cell layers [36]. This leads to scarring in the retinal segments. Thermal scarring enlarges progressively up to 300% and can cause significant vision loss if the fovea region is involved [37]. Thus, it is important to minimize the thermal effect during laser irradiations.

If pulse laser irradiations do not undergo photochemical or phase-transition processes, the linearly absorbed energy by the target tissue is entirely converted to the temperature rise. Since the focus of the current work is on femtosecond laser interaction within retinal or corneal tissues, we consider a situation where excessive laser energy, sufficiently higher than the threshold, is applied and absorbed in a small focal volume under single pulse configurations. A single shot of a fs-laser could be the entire source of heat. The temperature rise could be determined by calculating the volumetric energy density gained by the plasma during the laser pulse irradiations [38]. Under adiabatic conditions, the local temperature rise ($\Delta T(r)$) at an arbitrary location (x), is directly related to the local volumetric energy density ε (x), as.

$$\Delta T(x) = \frac{\varepsilon\,(x)}{\rho C_v} \tag{6}$$

where ρ is the tissue density and C_v is the specific heat capacity per unit volume [19]. In case of the absence of photochemical or phase-transition, the absorbed laser energy is subjected to undergo spatial redistribution by thermal diffusion leads to collateral damage in adjacent focal volume [39]. The heated volume is a layer of tissue, where penetration length is inversely proportional to an absorption coefficient ($1/\mu_a$) and the thermal diffusion time, (t_d), is expressed as:

$$t_d = \frac{1}{\kappa \mu_a^2} \tag{7}$$

where κ is the thermal diffusivity. In order to achieve thermal confinement, the ratio of pulse duration (t_p) to thermal diffusion time (t_d) should be less than or equivalent to 1 ($t_p/t_d \leq 1$). **Figure 9** shows the normalized temperature profiles in

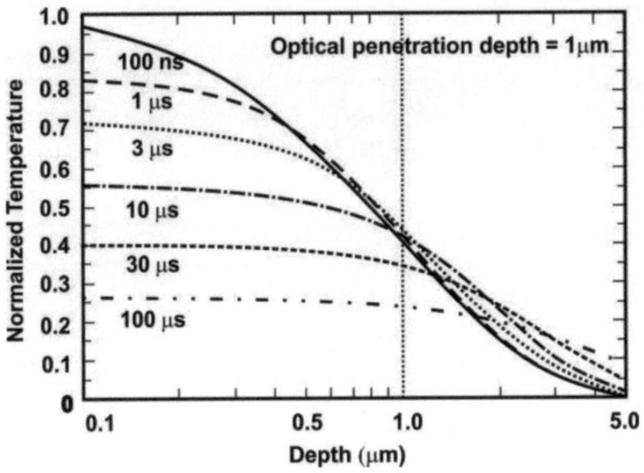

Figure 9.
Normalized temperature profiles in water immediately after the laser irradiation with constant irradiance and optical penetration depth (1 μm) at various pulse durations [19, 32].

water followed by laser irradiations with fixed radiant exposure and optical penetration depth (1 μm) for various pulse durations. At typical t_p < 3 μs, the temperature profile is confined to the diffraction-limited volume. However, for $t_p \geq 10$ μs, the thermal diffusion has spatially redistributed the energy over the larger volume. The peak temperature also reduced significantly within the sample over time [19, 32]. Thus, pulse durations play a vital role in thermal confinement to achieve localized ablations in the target tissue.

Meanwhile, collagen fibrils and water in corneal stroma often the main chromophores contribute to the absorption of IR and UV irradiations. Collagen fibrils in corneal stroma possess microscopic tissue structures in range of 30 nm in diameter, with corresponding thermal confinement times of 6.3 ns [40]. For micro-scale thermal confinement laser pulses of picoseconds or femtosecond, duration should be employed.

5.4 Plasma-induced ablation and photodisruption

Laser-induced optical breakdown occurs in biological tissues when the applied laser intensity exceeds 10^{11} W/cm^2. In case, an intense laser pulse can excite a large number of electrons and generate plasma, which causes vaporization of the materials [10–14, 18]. When a highly intense ultrashort laser pulse (<10 ps) was focused into the tissue, the associated energy density is high enough to induce nonlinear absorption of laser energy through multi-photon, tunneling, and avalanche ionization [10, 16, 19, 20]. Thus, producing micrometer-sized highly excited plasma in the vicinity of the focal volume. Moreover, within the stipulated pulse duration time, the temperature of the laser-induced plasma can reach several thousand Kelvin.

During the process of optical breakdown, plasma generation allows energy deposition in pigmented, as well as weakly absorbing tissues, such as cornea, lens, or retina. Laser-induced plasma will serve as an absorber of photon energy which leads to an increment in absorption coefficient [18, 32]. Thus, by means of plasma-induced ablations, clean removal of tissue without evidence of thermal or mechanical damage can be achieved even in transparent tissues.

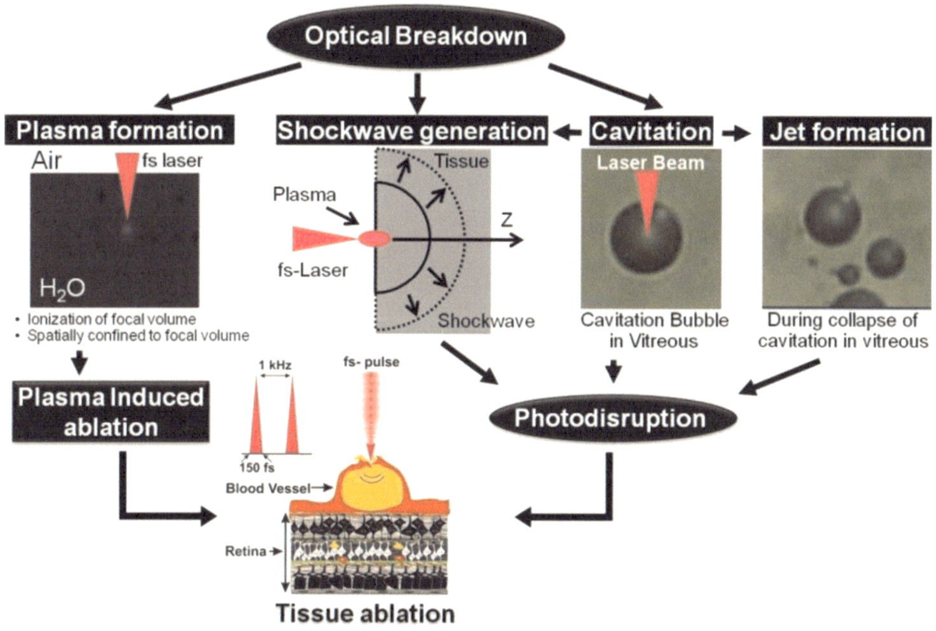

Figure 10.
Schematic of the physical effects associated with optical breakdown [32].

Alongside plasma generation, the movement of energetic free-electrons in plasma outward from the focal volume results in secondary effects, such as shockwave generation [18, 19, 32]. The shockwave leaves the plasma boundary at supersonic velocity and then slows down to the speed of sound, refers to as acoustic wave generation. The processes associated with an optical breakdown on the interaction of fs–laser pulse with materials were shown in **Figure 10**. Lastly, if energy density still lasts at the focal volume, it forms a cavitation bubble, which performs several oscillations of expansions and collapses within a period of few hundred microseconds under external pressure [41]. **Figure 11** illustrates the scheme of physical processes associated with optical breakdown. Plasma formation followed by mechanical effects could be referred as "*photodisruption*." It could be distinguished from plasma-induced ablations on the basis of employed laser energy density.

In plasma-induced ablations, the plasma formation vaporizes tissue at focal volume with lower threshold energies, providing a localized surgical effect. While, at higher energy densities, photo-disruption results in shock wave expansion, the expansion of the cavitation bubble, as well as heat diffusion after thermal equilibration, Thus causing all unwanted collateral damage to the surrounding tissue and limiting the surgical precision [10, 18, 19, 42, 43]. Thus, applied energy density, which is defined as energy per unit area, plays a crucial role to limit the collateral damage. The fluence (J/cm^2) employed in the sample must be minimized while still maintaining a sufficiently higher intensity to produce photodisruption through plasma formation.

Recent studies on plasma formation in water revealed some vital trends for pulse durations dependence while inducing optical breakdown. First, on reduction of pulse duration from 100 ns to 100 fs, the irradiance threshold for breakdown increases by ~1000 folds, but the radiant exposure threshold decreases from $\sim 10^3$ to $\sim 1\ J/cm^2$ [42, 43]. For nanosecond pulses, seed electrons for plasma formation

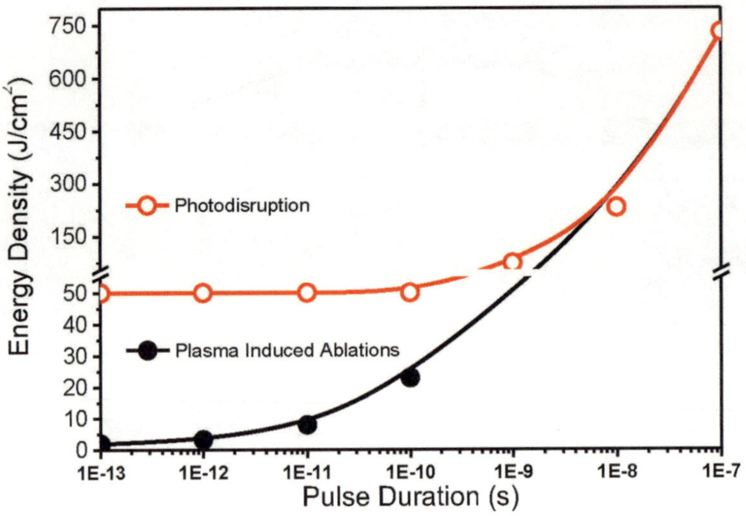

Figure 11.
Plasma-induced ablation and photodisruption distinction on the basis of energy density in cornea tissue [32].

were generated through thermionic emission which requires higher peak intensity. Rather, higher intensity of laser field generated by picosecond and femtosecond laser pulses can cause multi-photon ionization, which supplies the seed-free electron needed to start plasma generation at much lower peak intensities. Thus, the difference in radiant exposure threshold comes from the mode of free-electron generation or the initiation of plasma generation. As a result, nanosecond pulse lasers need higher threshold energy for plasma generation as compared to femtosecond and picosecond lasers. Second, the plasma transmission is small for ns-pulses, increases considerably for picoseconds pulses with maxima around 3 ps, and decreases again for fs-pulses [43, 44]. Lastly, the plasma energy density is more than one-fold smaller with fs-pulses than with ns-pulses [42–44]. Considering the facts, fs lasers could be employed as a microsurgical tool for precise ablations of biological tissues.

6. Ablation of biomatertials

6.1 Retinal blood vessels

In this section of the chapter, the ablation thresholds for retinal blood vessels of porcine have been studied as a function of vessel wall thickness. Vessel wall thickness, as well as the lumen diameter of porcine retinal blood vessels, is gradually decreasing while moving from the optic disc to the peripheral region of the retina (**Table 1**) and it was classified as primary, secondary, tertiary, and quaternary vessels as shown in **Figure 12(a)–(e)**.

6.1.1 Experimental technique

The experimental setup for fs-laser ablation of a porcine retina and as well as its blood vessels has been discussed in Ref. [45]. The laser system consists of a

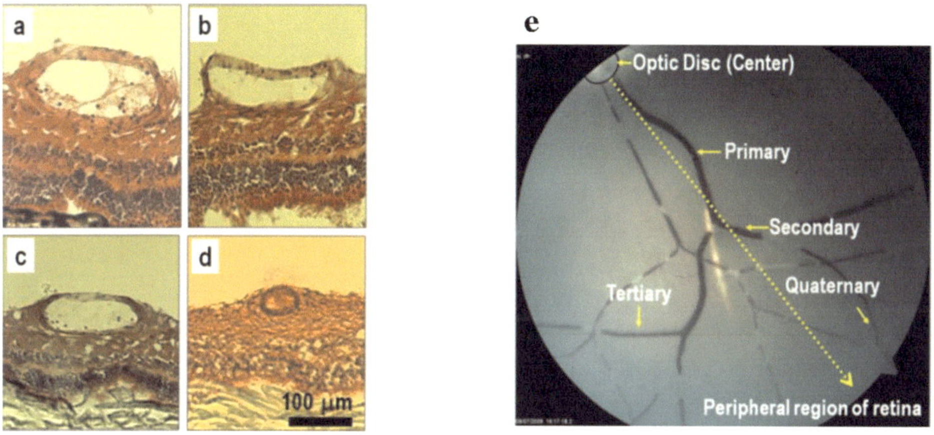

Figure 12.
H and E stained cryosections of retina blood vessels from the central region (optic nerve) toward the peripheral region in the porcine eye. (a) Primary (b) secondary (c) tertiary, and (d) quaternary. Scale bar is 100 µm. (e) Fundus image of porcine retina, indicating organization of retinal blood vessels from the optic disc "center" to the peripheral region [32].

regenerative amplified Ti: Sapphire (λ = 810 nm) laser, having a pulse width of 150-fs pulse at a repetition rate of 1 kHz (Quantronix, USA). The variable neutral density filter (Sigma, Kochi, Japan) is used for controlling the laser power. An objective lens (20X, Nikkon, Japan), was employed to focus the laser beam on the substrate surface, having a numerical aperture (N.A.) of 0.4. The diffraction-limited spot size at focus is given by.

$$D_{\min} = \frac{4\lambda}{\pi NA} \tag{8}$$

where λ is the wavelength (810 nm). The current optics used in this work result in a laser spot size of 2.6 µm in diameter. In general, the actual spot size is larger than the calculated value due to discrepancies in alignment.

To estimate the vessel ablation threshold, a series of laser fluence from 0.4 to 28 J/cm^2 for quaternary, from 0.7 to 43 J/cm^2 for tertiary, from 0.7 to 71 J/cm^2 for secondary, and from 1.4 to 99 J/cm^2 for primary retinal blood vessel was employed. For 60 different porcine eyeballs, 20–25 laser treatments were conducted at periodic intervals of 100 µm on each type of blood vessel. All the experiments were conducted under a single-shot configuration. Post-treatment of blood vessels, the histological analysis was carried out for determining the probability (%) of damage and probability (%) of vessels perforation for overlaid Inner Limiting Membrane [ILM] and the retinal blood vessels, respectively. The ILM could be defined as a thin layer membrane about ~6 um thick over the retinal vessels. Prior to the ablation of the blood vessel lumen, ILM must be ablated [32, 46].

6.1.2 Laser ablations of retinal blood vessels

Femtosecond laser irradiations were employed on all four categorized blood vessels, where the threshold for ILM ablation and the vessel perforation was determined from the H & E images taken off the mapped laser lesions. Later on, the wall thickness and determining thresholds were correlated to elucidate the relationship.

Organization level	Vessel thickness (μm)	Vessel diameter (μm)
Primary	10.01 ± 0.28	146.12 ± 10.55
Secondary	7.84 ± 0.11	118.79 ± 7.62
Tertiary	6.64 ± 0.71	93.11 ± 4.93
Quaternary	5.76 ± 0.86	51.42 ± 7.32

Table 1.
Vessel wall thickness and lumen diameter of porcine retinal blood vessels [32].

By changing the laser fluence from 1.4 to 99 J/cm^2, we produce a series of lesions on the primary blood vessel walls with a single fs-laser pulse. Considering the gradual decrease in total wall thickness of secondary, tertiary, and quaternary blood vessels, the lowest fluence at which no ablations on ILM was found to be 0.4 J/cm^2. However, the fluence at which the 100% probability of vessel perforation achieved was declined with the decrease in organization hierarchy from primary (99 J/cm^2), secondary (71 J/cm^2), tertiary (43 J/cm^2) to quaternary (28 J/cm^2) level. In all levels of blood vessels, with the progressive increase in laser fluence, the probability of ablation of the ILM and the vessel perforation show a monotonic increase. **Figure 13** exhibits the histological images of the sectioned slices of vessels after successive laser irradiations (a) primary, (b) secondary (c) tertiary, and (d) quaternary. As the size of the quaternary vessels is quite small, it is difficult to locate the laser lesions on the surface of blood vessels, therefore, 40× objective is used to map the laser lesions.

On analysis of histological sections, no apparent ablation of the ILM and no blood outflow were observed for the laser fluence less than 1.4 J/cm^2 in the case of secondary and tertiary blood vessels, whereas this value was further decreased to 0.5 J/cm^2 for quaternary blood vessels.

As per the analysis of cryosections, the lateral damage of the retinal blood vessels on fs-laser exposure is limited to 20 μm range at lower laser fluences. First apparent optoperforation of secondary and tertiary blood vessels was observed at the laser fluence of 3.6 J/cm^2 (**Figure 13(b)**-C and **(c)**-C), and on quaternary blood vessels it was noticed at 0.7 J/cm^2. Even though the blood vessel size is tens of micrometers, due to the precise focusing of fs-laser irradiations any damage to adjacent and underlying retinal-tissue is avoided. Meanwhile, on increasing the laser fluence to 14 J/cm^2 for secondary, 7.1 J/cm^2 for tertiary, and 3.6 J/cm^2 for quaternary blood vessels, single-pulse laser irradiation induces complete optoperforation of the blood vessel wall.

The cryosections of the blood vessels irradiated with a single-shot ultrafast laser were grouped into three types of lesions, including no change, ablation only at the ILM, and optoperforation of blood vessel walls. On this basis, correlation statistics of these different types of lesions are shown in **Figure 14** as a function of laser fluence. The perforation probability of blood vessels increased in the fluence range of 0.4–99 J/cm^2. The ablation threshold of all the blood vessels has been illustrated in **Table 2**. We determined the laser fluence to ablate the ILM layer over the blood vessels. It was found to be in the range of 1.4–3.6 J/cm^2 for primary, 0.7–1.4 J/cm^2 for secondary and tertiary, and 0.4–0.5 J/cm^2 for quaternary blood vessels. However, fs-laser-assisted perforation of blood vessels was achieved at higher laser fluence. Vessel perforation of secondary, tertiary, and quaternary vessels could be achieved with a fluence between 2.5–3.6 J/cm^2, 1.4–2.5 J/cm^2, and 0.5–1.4 J/cm^2, respectively.

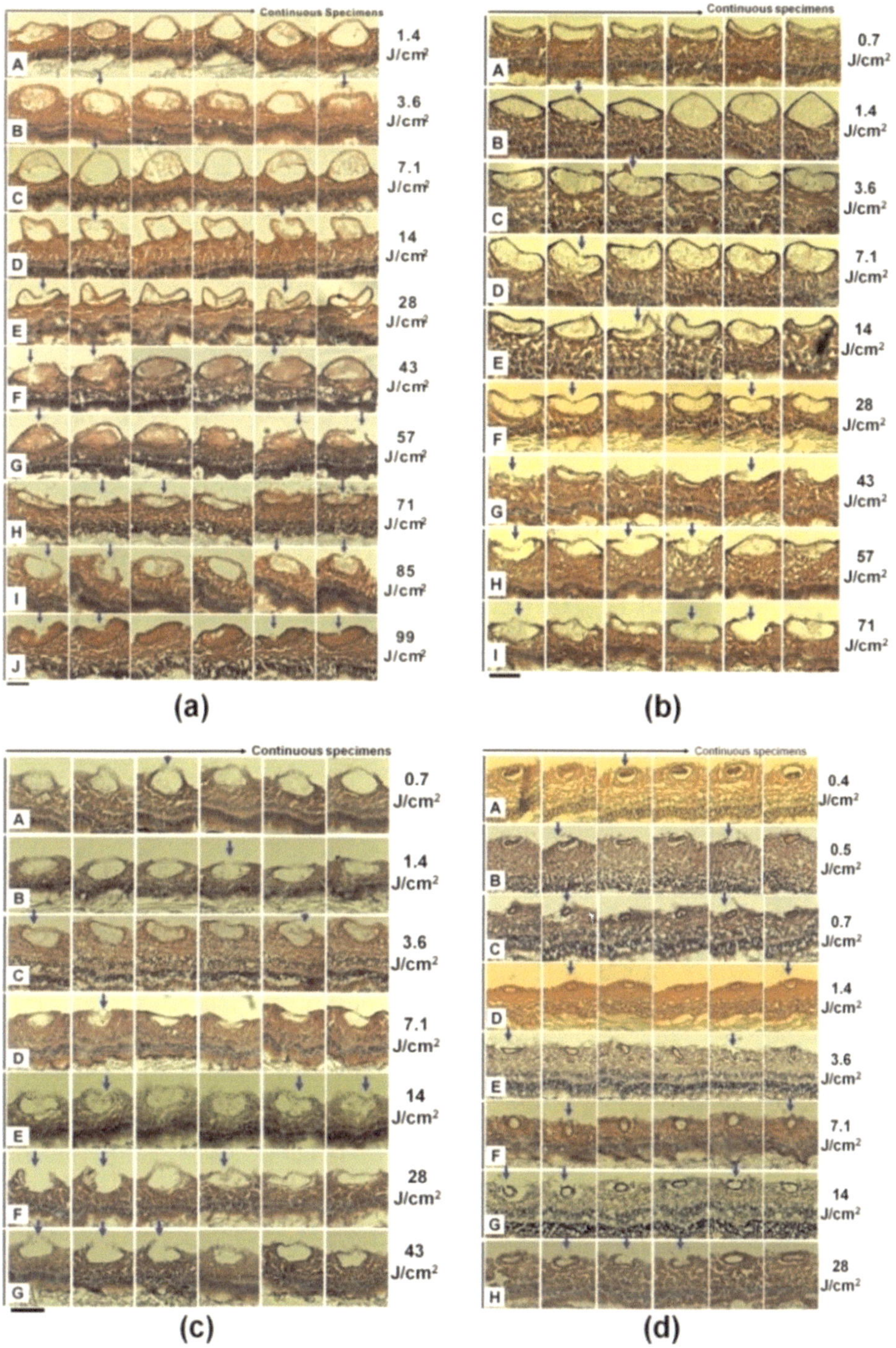

Figure 13.
Sequential H & E stained cryosections of retinal blood vessels after single fs- laser irradiations. (t = 20 μm). (a) Primary (b) secondary (c) tertiary, and (d) quaternary retinal blood vessels. Overall incident laser fluence is raised from 0.4 to 99 J/cm² or until 100% vessel perforations were achieved at each organization level. Scale bar is 100 μm [32, 46].

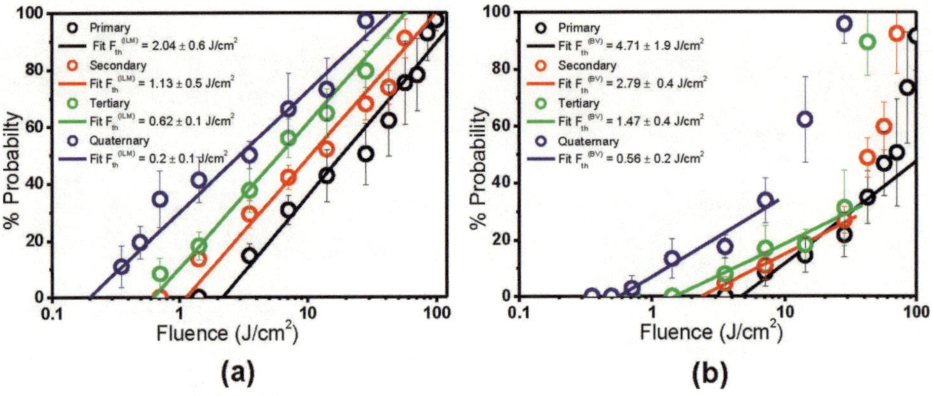

Figure 14.
(a) Linear plot of percent probability of damage for inner limiting membrane (ILM) and (b) vessel perforation as a function of the laser fluence for primary, secondary, tertiary, and quaternary blood vessels. With a progressive increase in the laser fluence, the percent probability of the blood vessel perforation shows a monotonic increase. The lines represent the extrapolation of the probability of damage to determine the ILM ablation threshold and vessel perforation threshold [32].

Organization level	ILM ablation threshold (J/cm^2)	Vessel perforation (J/cm^2)
Quaternary	0.20 ± 0.1	0.56 ± 0.2
Tertiary	0.62 ± 0.1	1.47 ± 0.4
Secondary	1.13 ± 0.5	2.79 ± 0.4
Primary	2.04 ± 0.5	4.71 ± 1.9

Table 2.
ILM ablation and vessel perforation threshold for retinal blood vessels (porcine).

It is the nonlinear nature of ultrafast laser-tissue interaction that leads to the three-dimensional submicron confinement of the laser absorption below the surface. The high peak intensities of ultra-short laser pulses provide a high flux of photons that could be nonlinearly absorbed by the electrons. The ultrafast duration of the absorption process leads to a rapid and efficient plasma generation where the beam is focused. Therefore, pulse energies as minimum as a few nano-joules (nJ) are sufficient for ablation of sub-cellular structures when the beam is tightly focused to submicron size [19, 47–49].

Comparative analysis of blood vessel optoperforation thresholds and ILM ablation thresholds for various blood vessels shows an increment with the increase in the wall thickness and lumen diameter. This provides an idea to selectively operate the blood vessels while discriminating on the basis of size as well as wall thickness without any apparent collateral damage to the underlying cell layers.

6.1.3 Corneal vessel ablation

Near-Infrared (NIR) femtosecond (fs) laser pulses focused into a transparent cornea allow surgery on neovascular structures with minimal collateral damage following the phenomenon of nonlinear multi-photon absorption. The fundamental output from a regenerative amplified Ti-sapphirelaser with λ = 810 nm, having a pulse width of 150 fs and a repetition rate of 1 kHz (Libra, Coherent Inc., USA) was focused into the rat corneal stroma by an ophthalmoscope lens (focal length = 36 mm, Carl Zeiss Inc., Germany). The numerical aperture (N.A) was 0.16. The laser spot was circular with a spot diameter of about 7.6 μm (measured at $1/e^2$ in intensity) [50].

The system was indigenously made that includes a software-controlled laser aiming equipment with an xy-galvano scanner to track pre-assigned targets visualized in optical images of the rat cornea. A schematic diagram of the experimental setup is illustrated in **Figure 15**, where under anesthetic conditions, the rat body was placed on a motorized XYZ translation stage, used to manipulate the target to expose a fresh area of tissue at each laser scan.

The minimal visible laser (MVL) lesion threshold was estimated for corneal neovascularizations (*abnormal blood vessels grown under adverse conditions in the avascular cornea*) by varying the laser fluence from 2.2 to 8.6 J/cm^2. The area of a scan was 150 × 150 μm, and the number of incident laser pulses (about 400) was kept constant [50].

The MVL lesion ablation threshold over the abnormal corneal blood vessels referred to as "neovascular structures" was estimated from high-resolution CCD images captured before and after the fs-laser exposure (**Figure 16**) Any noticeable or

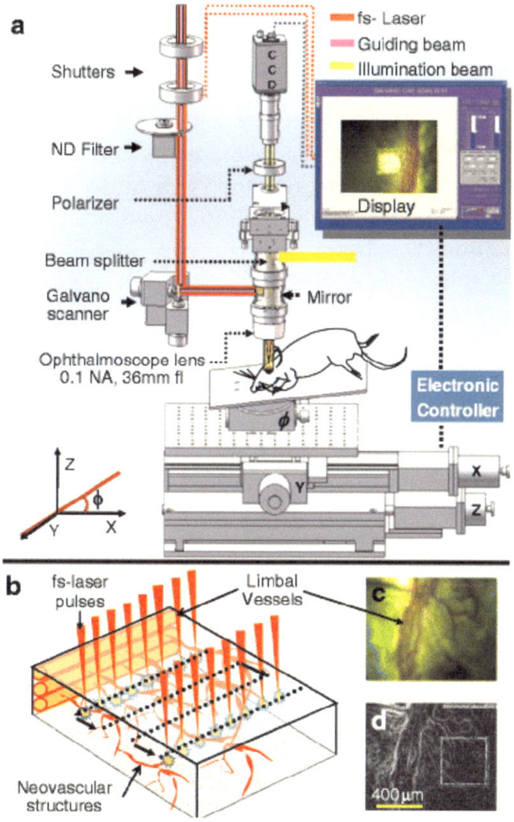

Figure 15.
(a) Schematic diagram of fs laser-assisted corneal neovascularization treatment system. (b) Schematic representation of scanning pattern of laser pulses into the corneal stroma. (c, d) Optical and transformed image of rat cornea captured before exposure of fs laser irradiation. Scale bar *is 400 μm [50].*

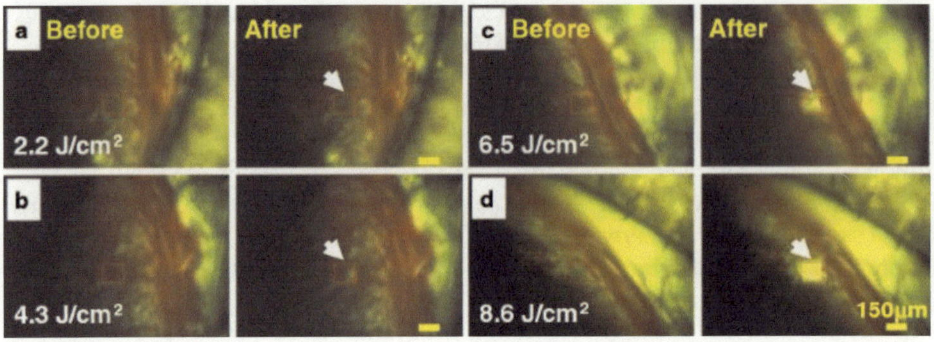

Figure 16.
Sequential CCD images of Norway brown rat cornea before and after the exposure of fs-laser irradiation where the laser fluence is 2.2 J/cm² (a), 4.3 J/cm² (b), 6.5 J/cm² (c), and 8.6 J/cm² (d). The scan area indicated by red-colored squares is 150 × 150 µm. Scale bar is 150 µm.

identifiable changes observed on neovascular structures under the high-resolution microscope in comparison to the neighboring intrastromal region immediately after the laser exposure were classified as an indication of damage.

At the fluence of 4.3 J/cm^2, the first visible, detectable lesion was found (**Figure 18b**) and is referred to as the minimal visible laser (MVL) lesion threshold. There was no clear damage found on either the intrastromal region or the neovascular structures at fluences <2.2 J/cm^2. When the set laser fluence was increased to 6.5 J/cm^2 or more, the size of the lesions also increased such that it covered the entire laser scanning area. For currently used femtosecond exposures, the laser pulse duration is shorter than the electron cooling and recombination times [19, 51, 52]. Thus, the minimal energy is nonlinearly absorbed during the pulse into the focused portion of tissues. However, the time scale of absorption is much shorter than both the thermal diffusion and shockwave propagation times. This might lead to localized photodisruption effects and subsequent reduction of stromal damage within the vicinity of laser focus [19, 51, 52]. fs ultrashort pulsedlasers for enclosure of corneal neovascularization in the presence of ICG at 3.8 J/cm^2 was employed by Sawa et al. in 2004 [53]. The MVL threshold values determined in the current study are in good agreement with previous reports [45, 53, 54], despite that no dye or photodynamic chemical agents were applied during the procedures.

6.2 Laser ablation of viscoelastic ultrathin spider silk fibers

We estimated a single-shot response of spider silk fiber of about 2–3 µm in diameter (**Figure 17**) [20, 55]. It is positioned at the focus using 100 x objective (NA: 0.9) with an estimated beam spot size around 1 micron. SEM micrographs illustrate the single-pulse ablation of silk fiber at 0.1 µJ and 0.3 µJ pulse energies. The estimated ablation threshold (F_{th}) for spider silk is about 0.05 µJ (1 J/cm^2). A slightly lower ablation threshold (0.6 J/cm^2) for silkworm silk is reported for different laser parameters with 230 fs pulses at the central wavelength of 515 nm wavelength [56]. This could be due to differences in the laser parameters, as well as the material properties [19, 20, 57]. F_{th} is found to be significantly higher than the ablation threshold for the crystalline Si, it could be attributed to the nonabsorbing nature of silk sample for incident laser wavelength, that is, 800 nm [19, 41]. By compensating for the dispersion of the fs-pulse, it

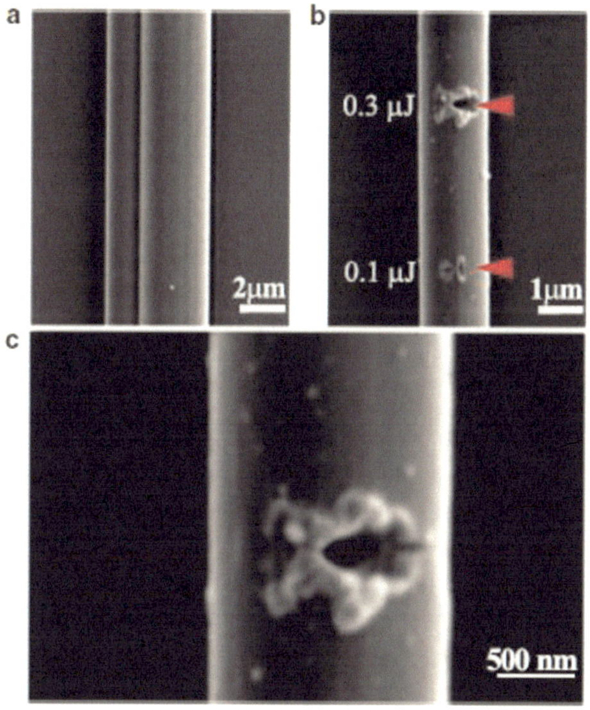

Figure 17.
(a) Native spider silk. (b) Illustrating ablation of silk fiber at different pulse energies 0.3 and 0.1 μJ, and (c) magnified SEM micrograph illustrating nanoscale ablation of spider silk microfiber at 0.3 μJ [55].

should be possible to further reduce the precision of ablation and may open a route to create nano-patterns on silk for various potential applications.

7. Electron beam ablations

Electron beams are used frequently for sterilization, as well as melting and welding metals. A novel e-beam-based material processing technology [58] called electron beam ablation has evolved to perform various surface functionalities. The EBA technique involves an e-beam deflected rapidly over a substrate surface to displace the material in a precisely controlled manner. This results in a textured surface containing an array of protrusions above the original surface and a corresponding array of cavities in the substrate. This process makes it possible to create custom-made, complex surface structures in various materials. The EBA is a complex process that comprises heating, phase change, and removal of a fine fraction from the target surface.

The precursor to this novel material processing technology is e-beam texturing [59]. In this process, the high-energy e-beam flux initially melts and then vaporizes the substrate material. The resulting vapor pressure then leads to the expulsion of molten material to the periphery of the hole. The electromagnetic coils focus the e-beam and then deflect it over the substrate material, leading to a rapid and controlled process. The typical processing speeds are 500–5000 holes per second, and the substrate surface may exhibit re-entrant features. Nowadays, another novel sculpting technology [59] has come into the picture to replace this texturing process. In this latest advancement, once a molten pool of material has been created, the beam is translated sideways. Under the combined effect of surface tension and vapor pressure, the material from the hole piles up behind the beam, as shown in **Figure 18**. This process is repeated several times at the same or overlapping sites to grow protrusions up to several millimeters accompanied by corresponding intrusions or holes. This technology makes it possible to grow a series of protrusions simultaneously across a substrate.

Careful control of the e-beam process parameters, such as beam accelerating potential, beam current and focus, the pattern design, and precision deflection movements allow for creating a range of different surfaces. These include spikes with a high aspect ratio, burr-free holes, channels, blades, swirls, and networks. It is possible to vary the size, shape, angle of incidence, and distribution of the features to produce customized surfaces within any pattern. Protrusions with dimensions ranging from tens of microns to several millimeters have been created successfully. Moreover, this technology has the flexibility to create a variety of structures, many of which may be impossible to produce using any other processing route. It is performed under a vacuum, thereby avoiding surface contamination. The technology has many potential applications, such as coating of thin films, fabrication of cardiovascular stents, and composite prosthetics. The pulsed-electron beam ablation technique for coating of thin films is described here.

Pulsed-electron beam ablation (PEBA) is an alternative approach that offers numerous advantages over other techniques. For example, for the preparation of thin films, the PEBA approach involves low capital cost, reduced operation/maintenance costs, a small footprint, and relative safety—no toxic gases as in PLA or potential noxious by-products as in solvo thermal routes. PEBA can be a potential candidate for epitaxial growth of high-quality thin films due to significant cost performance advantages, congruent multicomponent film stoichiometry under optimum conditions, process stability, and the scale-up ability for industrial applications.

A typical PEBA system consists of a pulsed-electron beam generator (pulsed-electron beam source), a stainless-steel deposition chamber, target and substrate holders, target and substrate rotating motors, substrate heater, and vacuum system. **Figure 19** shows a simplified schematic of the main pulsed-electron beam chamber components. To operate in pulsed mode, the electron source for high currents is appropriately modified. The channel spark is considered as the most efficient transient hollow cathode (THC) configuration for the generation of electron beams. THC

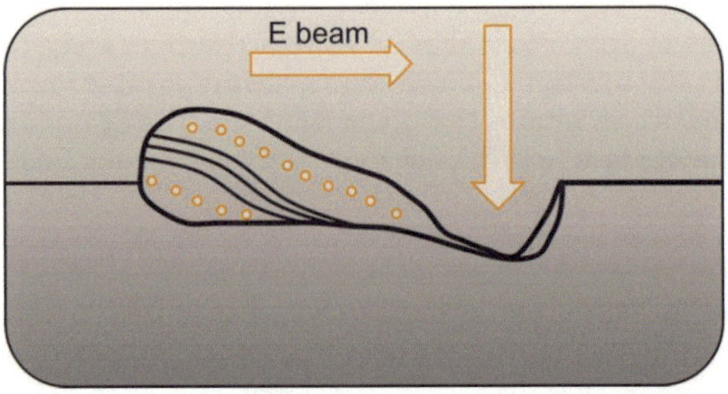

Figure 18.
Simplified schematic of a PEBA system.

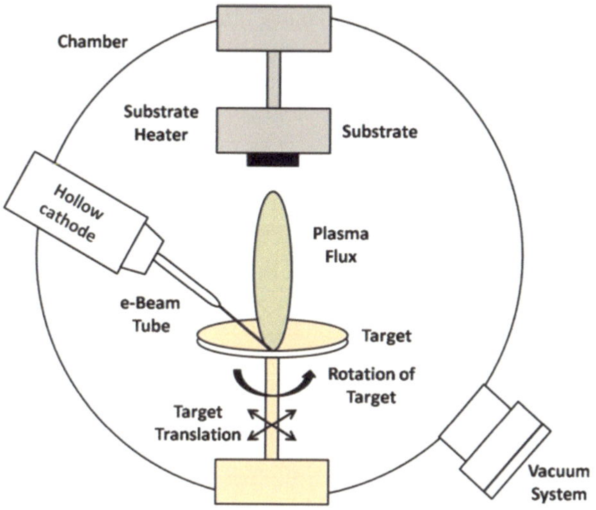

Figure 19.
Simplified schematic of a PEBA system.

acts as a low-pressure gas discharge electron source that can produce a focused electron beam with currents up to several kilo-amperes and a pulse duration of ~100 ns. Due to the high intensity of the generated beam passing through the deposition chamber, a self-pinch force is developed due to the ionization of background gas by the e-beam. This leads to the formation of a conducting plasma. This self-pinching feature enables the beam propagation, and thus target ablation, inside the deposition chamber without the application of external guiding fields [60, 61].

The triggering of the ablation process is caused by the pulsed-electron beam hitting the target surface. It is the pulsed nature of this technique that enables the confinement of energy within 1–2 μm thin subsurface of the target. Such confinement of energy leads to the instantaneous conversion of absorbed energy (energy of electrons) into the heat. The absorption of energy in such a minuscule volume of target material results in a jet evaporation at the target surface. This vapor cloud at the target surface continuously absorbs the majority of the pulsed-electron beam energy, gets ionized, and eventually forms a plasma. A pressure gradient in a direction perpendicular to the surface derives the expansion of the dense plasma which is seen as a plasma-jet (plume). The plasma-plume (high energy species), generated as a result of the interaction of pulsed-electron beam with the target surface, expands in the direction of the maximum pressure gradient at a velocity of about 10^4 m/s [62].

It is possible to manipulate ablation spot size, propagation of plume, and its characteristics by variations in the pulsed-electron energy, pressure and chemistry of background gas, and distance between channel spark tube tip and target surface. Plume characteristics and ablation spot size affect other deposition conditions significantly [63]. The production of thin films by PEBA is strongly influenced by the physical parameters in the plume, such as mass distribution, ion and atom velocity, and the angular distribution, of the plume species. Specifically, the homogeneity in film deposition on a substrate and the thickness distribution are determined by the plume shape that evolved during the expansion from the target surface to the substrate. These characteristics are also dependent on the target-substrate distance and the substrate size.

Subsequent to the formation of the plasma plume, the ejected high-energy species impinge on the substrate surface. These energetic species result in the sputtering of some of the surface atoms. A collision region is created between the incident flow and the sputtered atoms. Consequently, a thermalized region is formed that acts as a source for condensation of particles and hence causes the film to grow. Finally, a state of thermal equilibrium is reached, when the condensation rate is higher than the rate of removal of particles by sputtering (**Figure 19**).

8. Conclusions

Direct surface modification techniques using lasers or electron beams are pretty popular due to their reasonable cost and easy adaptation for small and large surfaces with complex geometry. The new generation of low-power femtosecond lasers with unique properties has enhanced the possibility of their use in many new exciting applications. Notably, for the surface modification of polymers, excimer lasers offer great potential. In addition, the medical industry is another sector where laser ablation and electron beam ablation play a key role. The medical industry has an ever-growing demand for the surfaces of implants and stents to enhance their performance. The technological advancements in e-beam ablation have resulted in the development of a novel surface processing technique capable of addressing these needs by rapidly producing bespoke surface features in a clean environment. It is being investigated for a range of medical applications.

Acknowledgements

The authors acknowledge the support from Electron Microscopy & Nanoscience Laboratory, Department of Soil Science, Punjab Agricultural University, Ludhiana. The authors also acknowledge Dr. Kamal P. Singh from the femtosecond laser lab, Physical Sciences, IISER Mohali , INDIA for his kind contribution in Z-scan and multiphoton analysis work on single fiber.

Conflict of interest

The authors declare no conflict of interest.

DOI: http://dx.doi.org/10.5772/intechopen.106198

Author details

Mehra S. Sidhu* and Nitish Dhingra
Department of Soil Science, Electron Microscopy and Nanoscience Laboratory, Punjab Agricultural University, Ludhiana, India

*Address all correspondence to: sidhums@pau.edu

References

[1] Iordanova E, Yankov G, Garasz K. Surface modification of different materials by fs-laser irradiation. Bulgarian Journal of Physics. 2017;**44**:133

[2] Schwibbert K, Menzel F, Epperlein N, Bonse J, Krüger J. Bacterial adhesion on femtosecond laser-modified polyethylene. Materials. 2019;**12**:3107

[3] Xue X, Lu L, Wang Z, Li Y, Guan YC. Improving tribological behavior of laser textured Ti-20Zr-10Nb-4Ta alloy with dimple surface. Materials Letters. 2021; **305**:130876

[4] Liu Y, Yuan G, Guo C, Ngo C-V, Li W. Femtosecond laser fabrication and chemical coating of anti-corrosion ethylene-glycol repellent aluminum surfaces. Materials Letters. 2022;**323**: 132562

[5] Jiao L, Chua ZY, Moon SK, Song J, Bi G, Zheng HY. Femtosecond laser produced hydrophobic hierarchical structures on additive manufacturing parts. Nanomaterials. 2018;**8**:601

[6] Jalil SA, Akram M, Bhat JA, Hayes JJ, Singh SC, Kabbash ME, et al. Creating superhydrophobic and antibacterial surfaces on gold by femtosecond laser pulses. Applied Surface Science. 2020; **506**:144952

[7] Sun H, Han M, Niemz MH, Bille JF. Femtosecond laser corneal ablation threshold: Dependence on tissue depth and laser pulse width. Lasers in Surgery and Medicine. 2007;**39**:654

[8] Olivié, G., Giguère, D.,Vidal, F., Ozaki, T., Kieffer, J. C., Nada, O., and. Brunette I., Wavelength dependence of femtosecond laser ablation threshold of corneal stroma, Optic Express 16, 4121 (2008).

[9] Vogel A, Noack J, Huttman G, Paltauf G. Mechanisms of femtosecond laser nanosurgery of cells and tissues. Applied Physics B. 2005;**81**:1015

[10] Perry MD, Stuart BC, Banks PS, Feit MD, Yanovsky V, Rubenchik AM. Ultrashort-pulse laser machining of dielectric materials. Journal of Applied Physics. 1999;**85**(9):6803-6810

[11] Hoerauf H, Brix A, Winkler J, Droege G, Winter C, Birngruber R, et al. Photoablation of inner limiting membrane and inner retinal layers using the erbium: YAG-laser: An in vitro study. Lasers in Surgery and Medicine. 2006;**38**: 52-61

[12] Yu PK, Miller J, Cringle JS, Yu D-Y. Experimental retinal ablation using a fourth-harmonic 266 nm laser coupled with an optical fiber probe. Investigative Ophthalmology & Visual Science. 2006; **47**:1587-1593

[13] Zysset B, Fujimoto JG, Deutsch TF. Time- resolved measurement of picosecond optical breakdown. Applied Physics B: Lasers and Optics. 1989;**48**(2): 139-147

[14] Perry MD, Campbell EM. Petawatt Laser Proposal, LLNL Internal Report –1992. Anaheim, CA: Conference on Lasers and Electro-Optics; 1993

[15] Du D, Liu X, Squier J, Korn G, Mourou G. Laser-induced breakdown by impact ionization in SiO_2 with pulse widths from 7 ns to 150 fs. Applied Physics Letters. 1994;**64**:3071-3073

[16] Stuart BC, Feit MD, Rubenchik AM, Shore BW, Perry MD. Laser- induced damage in dielectrics with nanosecond and sub picoseconds pulses. Physical Review Letters. 1995;**74**:2248-2251

[17] Pronko PP, VanRompay P, Horvath A, Loesel CF, Juhasz T, Liu X, et al. Avalanche ionization and dielectric breakdown in silicon with ultrafast laser pulses. Physical Review B. 1998;**58**: 2387-2390

[18] Niemz MH. Laser—Tissue Interactions. 3rd ed. USA: Springer Inc; 2007

[19] Vogel A, Venugopalan V. Mechanisms of pulsed laser ablation of biological tissues. Chemical Reviews. 2003, 2003;**103**:577-644

[20] Sidhu MS, Kumar B, Singh KP. The processing and heterostructing of silk with light. Nature Materials. 2017;**11**: 938-945

[21] Kaiser A, Rethfeld B, Vicanek M, Simon G. Microscopic processes in dielectrics under irradiation by subpicosecond laser pulses. Physical Review B. 2000;**61**:11437-11450

[22] Wong K, Vongehr S, Kresin VV. Work functions, ionization potentials, and in between: Scaling relations based on the image-charge model. Physical Review B. 2003;**67**:035406-035414

[23] Stuart BC, Feit MD, Herman S, Rubenchik AM, Shore BW, Perry MD. Nanosecond-to-femtosecond laser-induced breakdown in dielectrics. Physical Review B. 1996;**53**:1749-1761

[24] Il'insky YA, Keldysch KA. Electromagnetic Response of Material Media. New York, USA: Plennum; 1994

[25] Liu X, Du D, Mourou G. Laser ablation and micromaching with ultrafast laser pulse. IEEE Journal of Quantum Electronics. 1997;**33**:1706-1716

[26] Manenkov AA. Ultimate laser intensities in transparent solids. Laser Physics. 1996;**6**(3):501-505

[27] Keldysh LV. Ionization In The Field Of A Strong Electromagnetic Wave. Soviet Physics JETP. 1965;**20**:1307-1314

[28] Shen N. Photo-Disruption of Biological Tissues Using Femtosecond Laser Pulses [PhD thesis]. Harvard University, USA; 2003

[29] Yust BG, Mimun LC, Sardar DK. Optical absorption and scattering of bovine cornea, lens and retina in near infrared region. Lasers in Medical Science. 2011;**27**(2):413-422

[30] Hanczyc P, Samoc M, Norden B. Multiphoton absorption in amyloid protein fibres. Nature Photonics. 2013;**7**: 969-972

[31] Applegate MB, Marelli B, Kaplan DL, Omenetto FG. Determination of multiphoton absorption of silk fibroin using the Z-scan technique. Optics Express. 2013;**21**:29637-29642

[32] Sidhu MS. Parametric study of femtosecond laser-assisted selective removal of neovascular structures in animal eye. University of Science and Technology. [PhD thesis]. 2014

[33] Brooks BJ, Ambati BK, Marcus DM, Ratanasit A. Photodynamic therapy for corneal neovascularisation and lipid degeneration. The British Journal of Ophthalmology. 2004;**88**:840

[34] Mita M, Kanamori T, Tomita M. Corneal heat scar caused by photodynamic therapy performed through an implanted corneal inlay. Journal of Cataract & Refractive Surgery. 2013;**39**:1768-1773

[35] Holzer MP, Solomon KD, Vroman DT, Sandoval HP, Margaron P, Kasper TJ, et al. Photodynamic therapy with verteporfin in a rabbit model of corneal

neovascularization. Investigative Ophthalmology & Visual Science. 2003; **44**(7):2954-2958

[36] Kulkarni GR. Laser-tissue interaction studies for medicine. The Bulletin of Materials Science. 1988;**11**: 239-244

[37] Schatz H, Madeira D, McDonald HR, Johnson RN. Progressive enlargement of laser scars following grid laser photocoagulation for diffuse diabetic macular edema. Archives of Ophthalmology. 1991;**109**(11):1549-1551

[38] Xu YG, Xu YS, Huang C, Feng Y, Li Y, Wang W. Development of a rabbit corneal equivalent using an acellular corneal matrix of a porcine substrate. Molecular Vision. 2008;**14**:2180-2218

[39] Welch AJ, van Gemert MJC. Optical Thermal Response of Laser- Irradiated Tissue. New York, USA: Plenum Press; 1995

[40] Venugopalan V, Nishioka NS, Mikic′ BB. Thermodynamic response of soft biological tissues to pulsed infrared laser irradiation. Biophysical Journal. 1996;**70**: 2981-2993

[41] Schaffer CB, Nishimura N, Glezer EN, Kim AM-T, Mazur E. Dynamics of femtosecond laser-induced breakdown in water from femtoseconds to micro-seconds. Optics Express. 2002;**10**:196-203

[42] Kennedy PK, Boppart SA, Hammer DX, Rockwell BA, Noojin GD, Roach WP. A first-order model for computation of laser-induced breakdown thresholds in ocular and aqueous media. Part II: Comparison to experiment. IEEE Journal of Quantum Electronics. 1995;**31**:2250-2257

[43] Vogel A, Noack J, Nahen K, Theisen D, Busch S, Parlitz U, et al. Energy balance of optical breakdown in water at nanosecond to femtosecond time scales. Applied Physics B: Lasers and Optics. 1999;**68**:271-280

[44] Hammer DX, Frenz M, Jansen ED, Thomas RJ, Noojin GD, Rockwell BA, et al. Shielding effectiveness of laser-induced breakdown in water for pulse durations from 5 ns to 125 fs. Applied Optics. 1997;**36**:5630-5640

[45] Sidhu MS, Woo SY, Kim WK, Lee HS, Yahng JS, Kim KJ, et al. Optoperforations of retinal blood vessels in an intact porcine eye by using a femtosecond laser assisted microsurgery system. Journal of the Korean Physical Society. 2011;**58**(6):1605-1613

[46] Sidhu MS, Kim EK, Woo S-Y, Song MC, Jeoung SC, Park Y-I. Femtosecond-laser-assisted optoperforation of the primary retinal blood vessels and the retina tissue of porcine eyes. The Journal of the Korean Physical Society. 2009;**55**: 467-476

[47] Schaffer CB, Jamison AO, Garcia JF, Mazur E. Ultrafast Lasers: Technology and Applications. New York: Marcel Dekker Inc; 2002. p. 395

[48] Vogel A, Noack J, Huttman G, Paltauf G. Mechanism of femtosecond laser nanosurgery of cells and tissues. Applied Physics B. 2005;**81**(8):1015-1047

[49] Yoon TO, Shin HJ, Jeoung SC, Park YI. Formation of superhydrophobic poly (dimethysiloxane) by ultrafast laser-induced surface modification. Optics Express. 2008;**16**(17):12715-12725

[50] Sidhu MS, Choi MY, Woo SY, et al. Femtosecond laser-assisted selective reduction of neovascularization in rat cornea. Lasers in Medical Science. 2014; **29**:1417-1427

[51] Vogel A. Nonlinear absorption: Intraocular microsurgery and laser lithotripsy. Physics in Medicine and Biology. 1997;**42**(5):895-912

[52] Vogel A, Noack A, Nahen K, Theisen D, Birngruber R, Hammer DX, et al. Laser-induced breakdown in the eye at pulse durations from 80 ns to 100 fs. Proceedings of SPIE. 1998;**3255**:34-49

[53] Sawa M, Awazu K, Takahashi T, Sakaguchi H, Horiike H, Ohji M, et al. Application of femtosecond ultrashort pulse laser to photodynamic therapy mediated by indocyanine green. The British Journal of Ophthalmology. 2004; **88**:826-831

[54] Giguère D, Olivié G, Vidal F, Toetsch S, Girard G, Ozaki T, et al. Laser ablation threshold dependence on pulse duration for fused silica and corneal tissues: Experiments and modeling. Journal of the Optical Society of America. A. 2007;**24**: 1562-1568

[55] Sidhu MS, Singh KP. Ablation of silicon and ultrathin fibers using single femtosecond pulse. Indian Journal of Physics. 2019;**93**:1619. DOI: 10.1007/s12648-019-01419-5

[56] Vollrath F, Madsen B, Shao Z. The effect of spinning conditions on the mechanics of a spider's dragline silk. Proceedings of the Royal Society of London B. 2001;**268**:2339

[57] Preuss S, Demchuk A, Stuke M. Sub-picosecond UV laser ablation of metals. Applied Physics A: Materials Science & Processing. 1995;**61**:33

[58] B.G.I. Dance and E.J.C. Kellar (Inventors) Workpiece Structure Modification, Surfi-Sculpt, International Patent Publication Number WO 2004/028731 A1 Applicant: The Welding Institute.

[59] B.G.I. Dance (Inventor). 'Modulated Surface Modification,' International Patent Publication Number WO 2002/094497 A3, Applicant: The Welding Institute

[60] Buxton AL, Dance BGI. Surfi-Sculpt – revolutionary surface processing with an electron beam. In Proceedings of 4th International Surface Engineering Congress, 1–3 August, St. Paul, Minnesota, USA. 2005. pp. 107-110. ISBN: 0-87170-835-3

[61] Fernsler RF, Hubbard RF, Lampe M. Pinched Propagation of High-power, Pulsed Electron Beams for Welding and Materials Processing Applications. Washington DC: Naval Research Lab; 1994. pp. 1-59

[62] Nistor M, Gherendi F, Mandache NB. Fast imaging of ablation plasma produced by a pulsed electron beam. Journal of Transactions on Plasma Science. 2011;**39**:2800-2801

[63] Christen HM, Lee DF, List FA, Cook SW, Leonard KJ, Heatherly L, et al. Pulsed electron deposition of fluorine based precursors for $YBa_2Cu_3O_{7-x}$-coated conductors. Journal of Superconductor Science and Technology. 2005;**18**:1168